The Worth of Water

Liam McCarton • Sean O'Hogain • Anna Reid

The Worth of Water

Designing Climate Resilient Rainwater
Harvesting Systems

 Springer

Liam McCarton
School of Civil and Structural Engineering
Technological University Dublin
Dublin, Ireland

Sean O'Hogain
School of Civil and Structural Engineering
Technological University Dublin
Dublin, Ireland

Anna Reid
School of Civil and Structural Engineering
Technological University Dublin
Dublin, Ireland

ISBN 978-3-030-50607-0   ISBN 978-3-030-50605-6   (eBook)
https://doi.org/10.1007/978-3-030-50605-6

This Springer imprint is published by the registered company Springer Nature Switzerland AG
The registered company address is: Gewerbestrasse 11, 6330 Cham, Switzerland

Foreword

At the time of writing, the world faces two challenges to our continued existence on the Earth, the Covid19 pandemic and the global impacts of climate change. The worldwide response to both has been informed by science. Science provides evidence and advice, prompting governments to act and business and individuals to respond.

The worldwide adaptations we humans have made to combat Covid19 show how a shared global framework, informed by scientific evidence and advice, can motivate radical action. The Covid19 response proves that when faced with an existential threat to our survival, we humans can change how we act, how we think and how we interact. On the other hand, the pandemic appears to have pushed all discussion of climate change, and the corresponding individual lifestyle and governance changes required, down the priority list.

Numerous scientific reports have shown evidence of the impacts of global warming and of the necessary actions needed to keep within 1.5–2°C limits. The scientific approach to policy change to date has been to provide evidence and data to support an informed debate and trust our politicians, legislators and business leaders to act accordingly. The Covid19 response features government action and lifestyle interventions to combat the effects of the public health emergency. This has not been the case with climate change.

Low income countries across the world are characterised by low levels of daily income, lack of adequate housing, water and sanitation infrastructure, limited health capacities and weak governance. High income countries in the so-called developed world have not prioritised this cohort of our brothers and sisters who face the brunt of climate change. On the contrary, reduced development budgets from wealthy countries have accentuated and promoted divisions and inequalities in these countries. In terms of addressing climate change, middle and high income countries have turned inwards. The rich world which has reached a high level of income largely founded on the use of non-renewable resources is in effect pulling the ladder up after it and leaving low income countries to fend for themselves.

The utilisation and distribution of the planetary resources is not shared equally. It is estimated that humanity needs 1.7 Earths to meet present consumption, with some high income countries needing the equivalent of 4 Earths to sustain their current lifestyle. In the face of this, there is a need for low income countries, and for the communities that make up these nations, to develop resilience. Resilience is the ability of a community to cope with destabilizing forces. In terms of water supply, sanitation and hygiene, this will mean increased reliance on self-supply. Decentralised technology in the areas of water supply and storage, used water treatment and water reuse will lead to a plug-and-play world where locally available materials will be adapted to each biosphere, with systems designed, maintained and operated by locals. A feature of these systems will be the safe and efficient harvesting of rainwater.

Water has been central to the ability of societies to blossom. The collapse of many ancient civilisations and cultures was due, in no small part, to local water-related shocks. The current challenges that the society is facing are more complex and dangerous, the threat occurring at both local and global level.

The worth of water is a combination of the value of water, as a life-support system, and the value in water, as the basis of economic life. The future will see the worth of water becoming central and will involve water being at the centre of all government and economic decisions. Recognising this is the foundation of combating the current pandemic and climate change and creating the conditions necessary for our planet to support all its children sustainably and within the planetary boundaries.

Thousands have lived without love, none without water.
W.H. Auden.

Preface

Living as we are in a post climate change era, adaption and innovation are terms that point the way toward the future. The world needs to adapt to new climatic patterns. These new climatic conditions threaten the very survival of some countries, they threaten our cities and communities and they threaten our ability to feed ourselves. Given the importance of water in the lives of all humans, we simply cannot live without it; it is no surprise that major changes are occurring in how we use water. The first chapter considers the historical concept of the Linear Economy of Water (LEW). It goes on to show how the increased emphasis on sustainability and global demographics led to the adoption of the Circular Economy of Water (CEW). This lays the emphasis on recycling and reuse of water and views water and all of the contents of domestic, industrial and agricultural waters as a resource. The CEW also considers multiple water sources including surface water and ground water together with alternative sources such as rainwater, brackish and saline water, brines, and used water. These sources of different water qualities are available for different water use applications. Therefore, the CEW supplies multiple waters for multiple purposes for multiple users. In contrast to the LEW, which focused on supplying potable water quality, the CEW deals with varying water qualities and the principle of multiple waters gives rise to the term "fit for purpose". This denotes that the water from the used water source can be utilised by an application without the used water requiring treatment.

Water and its properties are a function of its peculiar molecular structure. Chapter 2 presents this molecular structure of water and discusses how this structure influences the behaviour of the water molecule. The influence of polarity and hydrogen bonding on the three states of water and other properties of water such as density, cohesion, adhesion and surface tension are also discussed. The high heat energy required to warm water and its role in moderating the Earth's climate and the effect that increased temperatures have on the global system are examined. Finally, the unique properties of water as a universal solvent are presented.

Water quality, potable water production, water quality parameters and the treatment processes involved in potable water production are presented in Chap. 3. The means of removing these parameters, referred to as the removal mechanisms, are defined. There is also discussion on the type of treatment afforded and the parameters removed. The role of these removal mechanisms in a conventional rainwater harvesting system is presented. The location of the removal mechanisms in the rainwater harvesting system is emphasised.

The term "the worth of water" was developed to reflect the importance of the social life of water. Chapter 4 presents a technology portfolio of the worth of water from the ancient world to the present. In contrast to the present global situation, ancient cultures were aware of the worth of water and the importance to society of the social life of water. Case studies are presented from these cultures showing the technology and the importance of this technology to the very existence of these cultures. Ancient technologies from Iran, Peru and Yucatan are illustrated. The chapter goes on to examine examples of the worth of water and their suitability and importance to developing countries. A particular example of an innovative method of promoting water and water technologies in developing countries is presented in a village technology education centre. Finally, the chapter presents case studies of rainwater harvesting systems and their adaptation in the developed world and such modern countries as Singapore and South Korea.

Rainwater harvesting (rwh) technology captures water and conveys it to storage. Chapter 5 considers the water flows in a typical roof rwh system. The components of the rwh system are also discussed. A summary of the three types of roof rwh systems – direct system, gravity system and indirect systems – is presented as are some alternative methods of rwh systems which utilise ground-based systems for irrigation water or groundwater recharge. These include lined underground reservoir, contour ridges, contour stone bunding, terracing contour bunds, permeable rock dams, recharge pits/trenches and check dams.

The health implications of using rwh are an important consideration, and Chap. 6 investigates if domestic water supply systems supplied with harvested rainwater present an increased risk to health over systems supplied with potable mains water. Several studies are reviewed which conclude that the main risk to public health of mains-supplied hot water systems is the operation, maintenance, age, location and temperature of the system. Rainwater harvesting systems contain an inherent water treatment train which improves the water quality at different parts of the rwh system. Results from laboratory experiments conducted using a variety of water-related bacteria to determine the time required to reduce a bacterial population by 90% at a given temperature are compared with international studies. The results show that after 5 min of exposure at 60 and 55 °C, respectively, Salmonella and Pseudomonas aeruginosa and total viable count at 22 and 37 °C concentrations were reduced to zero. The conclusion from this chapter is that hot water systems supplied with harvested rainwater do not present an increased risk to health over hot water systems fed with mains water.

A design methodology to establish if rwh is feasible and to enable the designer to select the optimum rwh storage tank size is presented in Chap. 7. The design methodology consists of two stages. Stage 1 involves the designer establishing the hydraulic efficiency for a site. This involves calculating the rwh demand profile and rwh supply profile for this location. The supply coefficient is then established which is defined as the percentage of demand that the rwh system can supply for a selected time period. Stage 2 involves the designer calculating the optimum rwh storage tank volume. This can be established according to one of three methods, basic method, intermediate tabular method and/or a detailed daily storage model. A number of worked examples are detailed to explain the design methodology.

Chapter 8 presents a number of financial management tools which can be used to evaluate the economic performance of a rwh system. The simple payback period is used to determine the length of time a rwh investment reaches a break-even point. This is calculated by dividing the initial cost of the investment by the annual savings generated. The second tool discussed is the net present value (NPV) method, which takes into account the time value of money to evaluate the cost of ownership of a rwh system over its entire lifetime. The third technique discussed is where the cost of ownership over the lifetime of an asset is annualised and the equivalent annual cost (EAQ) is calculated. A series of worked examples are presented which compare the costs of a rwh system compared to various alternative public water supply options.

The Sustainable Development Goals (SDGs) and the importance of water within each of the goals is addressed in Chap. 9. The fundamental concept of sustainable development is discussed within the growing uncertainties of climate change and global population growth. The SDGs have the potential to form a template for development, and the importance of water within each goal is addressed. Finally, a critical review of the SDGs is presented. Some of the contradictions inherent within the SDG agenda are discussed and critically evaluated.

Build Solid Ground is a project funded by the European Union Development Education and Awareness (DEAR) programme and implemented by a consortium of 14 organisations from seven EU countries. The focus of the project is building critical understanding and active engagement for sustainable development goal 11, "sustainable cities and communities". Chapter 10 investigates the key components necessary to achieve a resilient city and community.

Chapter 11 presents a detailed discussion of the planetary boundaries. This is a framework, based on science, which identifies nine processes and systems that regulate the stability and resilience of the Earth system and the interactions of land, ocean, atmosphere and life that together provide conditions upon which our societies depend. The planetary boundaries (PBs) framework arises from the scientific evidence that the Earth is a single, complex, integrated system. Each of the planetary boundaries, climate change, biodiversity loss, stratospheric ozone depletion, ocean acidification, nutrient cycles, land system change, freshwater use, atmospheric aerosol loading and chemical pollution are discussed in detail in this chapter.

Human societies have been adapting to their environments throughout history by developing technology, cultures and livelihoods which are suited to local conditions. Climate change raises the possibility that existing societies will experience climatic shifts that previous experience has not prepared them. Chapter 12 discusses the role of rwh to enable society to adapt to climate change. It examines the fact that present infrastructure, with a maximum lifespan of 30 years, will not be the infrastructure of the future. Grey, green or hybrid infrastructure and the increasing role to be played by decentralised systems are examined. The role of rwh technology in future systems is compared to the role played by wind and solar energy resources. The chapter discusses that though rwh will not supply 100% of water demand, it can supply a substantial alternative source of water. The potential in developing countries for rwh technology to provide the opportunity to leapfrog capital-intensive water projects by going straight to decentralised water collection systems is stressed. The role of multiple waters in climate adaption is examined as is the concept of peak water. Finally, the potential for rwh technology to facilitate the rebalancing in society of the human technology environment nexus is addressed.

Liam McCarton
Sean O'Hogain
Anna Reid

Acknowledgements

Dedicated to Angie for a life full of adventure, energy and fun. This book has had help from many people and many organisations. Not all people were aware of this help and neither were some of the organisations. However, thanks go out to them all. To our families, Lyn, Jackie, Angie and Sadhbh. To Prof. Joan Garcia of Universitat Politecnica de Catalyuna, Albert Jansen and Victor Beumer from Utrecht, Alenka Mubi Zalaznik of Limnos in Slovenia and to all the team at Limnos. To the staff of Habitat for Humanity. To our partners in the EU DEAR Build Solid Ground Project. To Declan, Katie, Emma, Andrew and all at Engineers Without Borders Ireland. To Pat Kennedy and Noreen Layden. To the usual suspects in TU Dublin – John Turner, Richard Tobin, Aidan Dorgan, Stephen McCabe, Catherine Carson, Catherine McGarvey, Debbie McCarthy, John Donovan, Michelle O'Brien, Liz Darcy and Marek Rebow.

The authors acknowledge the financial support of the European Union and the Build Solid Ground DEAR project.

All graphical illustrations by Maia Thomas (maia_thomas@hotmail.com)

Introduction

Organisation of the Book

The book is presented in three parts. Chapters 1, 2, 3, and 4 discuss some of the scientific principles and core concept within the book. Chapters 5, 6, 7, and 8 assess some of the technical, economic and health aspects of rainwater harvesting systems. Chapters 9, 10, 11 and 12 then discuss rwh within the context of sustainable development and climate adaption.

Chapter 1 introduces the concepts of the linear economy of water (LEW), the circular economy of water (CEW), the worth of water and multiple waters. In contrast to the LEW, which focused on supplying potable water quality, the CEW deals with varying water qualities, and the principle of multiple waters gives rise to the concept that the use of the water governs the water quality. Chapter 2 presents the molecular structure of water and discusses how this structure influences the behaviour of the water molecule. Water quality, potable water production and water quality parameters and the treatment processes involved in potable water production are presented in Chap. 3. The term "the worth of water" was developed to reflect the importance of the social life of water. Chapter 4 presents a technology portfolio of the worth of water from the ancient world to the present. Chapter 5 considers the water flows in a typical roof rainwater harvesting (rwh) system. A summary of the three types of roof rwh systems – direct system, gravity system and indirect systems – is presented as are some alternative methods of rwh systems which utilise ground-based systems for irrigation water or groundwater recharge. The health implications of using rwh are an important consideration and Chap. 6 investigates if domestic water supply systems supplied with harvested rainwater present an increased risk to health over systems supplied with potable mains water. A design methodology to establish if rwh is feasible and to enable the designer to select the optimum rwh storage tank size is presented in Chap. 7. Chapter 8 presents a number of financial management tools which can be used to evaluate the economic performance of a rwh system. The Sustainable Development Goals (SDGs) and the importance of water within each of the goals is addressed in Chap. 9. Chapter 10 gives a brief overview of the objectives of the EU DEAR project "Build Solid Ground" and the components of resilient cities and communities. Chapter 11 presents a detailed discussion of the planetary boundaries. This is a framework, based on science, which identifies nine processes and systems that regulate the stability and resilience of the Earth system and the interactions of land, ocean, atmosphere and life that together provide conditions upon which our societies depend. Chapter 12 discusses the role of rwh to enable society to adapt to climate change.

Contents

List of Figures

List of Tables

Chapter 1
The Worth of Water

1.1 Introduction

The way we view water has changed over the last decade of the twentieth century and the first decades of the twenty-first century. Previously water was subject to a linear form of treatment where it was abstracted, treated before use to drinking water (potable) standards, used and treated again before finally being discharged back to the environment through ground or surface water. Water Engineering involved the identification and evaluation of raw water sources, with a view to using a water source that required minimal treatment to remove natural pollutants such as nitrates. This raw water was then impounded, sometimes in reservoirs or dams. Quality issues arose in depth of draw off, as spring and autumn turnovers in the impounded water gave rise to chemicals being dissolved in the water and requiring treatment. All impounded water was then treated in water treatment plants to potable water quality standards.

© Springer Nature Switzerland AG 2021
L. McCarton et al., *The Worth of Water*,
https://doi.org/10.1007/978-3-030-50605-6_1

This potable water required pipe networks to supply the water to the populace. Quality issues in these networks was a concern. This was due to the age of the networks which often gave rise to loss of water and, more importantly, ingress of untreated soil water, which oftentimes altered the water quality. Once used by the populace, this water, termed wastewater was seen as a potential source of disease. Domestic waste water contained human and food waste while industrial wastewater contained a range of inorganic and organic chemicals seen as harmful to the environment. Therefore, pipe networks were required to remove these wastes, efficiently and safely, from the point of use to a site where the water could be treated to a standard that allowed for discharge to the environment. In most cases this resulted in the discharge of water to a river, lake or the sea.

1.2   The Linear Economy of Water

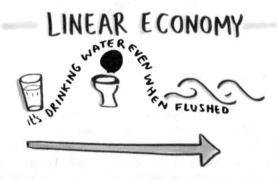

Stringent regulations were applied to the quality of water used for human consumption. Potable or Drinking Water Standards were introduced over the course of the twentieth century to ensure the health and safety of the population when consuming urban and rural water supplies. These standards laid down a series of physical, chemical and microbiological parameters which had to be met before water could be consumed by the general public. Stringent standards were also set down for wastewater being discharged to receiving waters. This resulted in the hierarchy of potable water standards, non-potable water standards and waste water standards. In the scheme of the Linear Economy of water, alternative sources of water, e. g. rainwater harvesting, were discouraged as they were not of sufficiently high quality

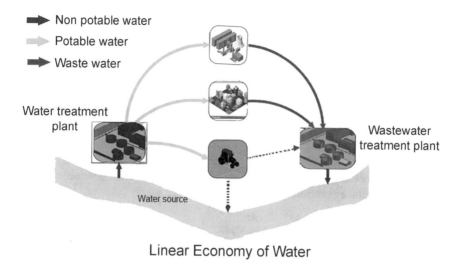

Fig. 1.1 The linear economy of water

to serve as potable waters. Further there existed the danger, particularly in the domestic situation, that these alternative sources could contaminate and compromise potable supplies, thus leading to outbreaks of disease. This linear economy of water, where all water is abstracted, treated to potable standards, used and then treated again to wastewater standard prior to discharge is shown in Fig. 1.1.

1.2.1 Leaving the Linear Economy of Water Behind

An increasing world population requires increasing water resources. The number of people in the world is increasing and they are also on the move. People are moving into urban areas and this is especially true of urbanisation along coastal areas. Projections show that by 2030, 60% of the world's population will be living near or in coastal regions. The water supplied to these coastal urban cities was historically supplied from inland areas where the original raw water sources were of a very high quality. The water was transported to the urban population centres, treated and supplied to the populace using a supply network. It was then used, in most domestic situations used only once, before being collected, by a different network to the supply network and being treated and discharged to the sea. Increased demand on resources together with the discharge of a treated effluent that could serve as a raw water source lead to this practice being seen as unsustainable.

This movement to the cities is also accompanied by radical changes in the world's weather patterns. Global warming increasingly impinges on rainfall patterns altering the amount of rainfall but also altering the severity of rainfall and its frequency. Storm events and flooding increasingly affected urban areas but so, unthinkably, does drought. The issue of resources utilisation and the sustainability of the way the world consumed these resources has become an issue. The term sustainability, (i.e. the efficient use of finite resources) begins to focus minds on the use of water.

Water and water use is becoming a focus for change. This is driven by the increasing effect global warming is having on the water cycle, the melting of glaciers, reduction in river flow and severe storm events that cost many lives and has resulted in disruption of daily life together with huge financial loss. The linear economy of water has become an outdated water management and supply concept.

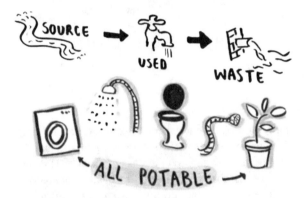

1.3   The Circular Economy of Water

The global water cycle is often referred to as a circle. Water falls on the Earth and is either evaporated as it runs over ground, collects in water bodies such as rivers, lakes and groundwater or falls into the sea. All water bodies give rise to evaporation, where the water is returned to the atmosphere where it can fall again. Water that flows to underground sources serves as groundwaters for use in agriculture or as the base flow into rivers or streams. Water that is captured as snow or in glaciers serves as the source for rivers that also take part in the cycle. The natural process operates as a circle, it generates no waste.

The **Circular Economy of Water (CEW)** is also a circle/cycle. It views water and its contents as a resource. This content may be water itself, as water is also a content in all water that is used, either for domestic, industrial or some agricultural processes. The CEW is waste free and also views used water (wastewater) as a resource. Used water (wastewater) can be treated by removing the dissolved or particulate matter in it potentially resulting in a potable water source. Other resources are also contained in used water, and these resources can range from heat, to minerals, to dissolved organics as well finite resource nutrients such as nitrogen and phosphorus which can be recovered and reused.

The circular economy of water has as its aim the maximisation of resource recovery from all water processes. It also promotes the recycling and reuse of water wherever possible. Taken together these two principles change the way we view water use. Instead of thinking in terms of discharge of effluents, to sewers, to treatment plants or to water bodies (receiving waters) we now consider used water (water plus resources) as a source of value, where the water and the contained resources are referred to as the value in water.

1.3.1 From Wastewater to Used Water

The CEW does not include or use the conventional term wastewater. The term waste, applied to wastewater or solid waste denotes no longer usable or useful. Waste has to be removed and treated or stored. In the CEW there is no waste as the used water and the resources within, serve as raw material for industrial processes, agricultural processes and indeed domestic use. Water is not considered in a linear fashion with the need to remove pollutants and then discharge it, but is seen as existing as multiple waters with multiple uses.

1.4   Multiple Waters, Multiple Uses, Multiple Qualities

In the CEW multiple water sources refers to not only surface and ground waters but also to alternative sources such as rainwater, brackish and saline water, brines and used water. Thus these sources are seen not only as sources of waters of different qualities but also as sources of waters with different applications. The old linear economy of water concept of potable water and wastewater is replaced by the concept of multiple water for multiple uses. The CEW promotes water management processes that will provide and allocate water by storing, treating and distributing the right water for the right purpose to the right users. This will see rainwater harvested by green infrastructure in cities and communities and made available for use. New local loops and decentralised water treatment systems will ensure that used water from apartment blocks or living quarters can be recycled and reused, and options for extracting, valorising and using nutrients in the used water streams for fertilizers in integrated urban natural environments will also evolve. As a result different alternative water sources and qualities (fresh ground and surface water, rainwater, brackish water, saline water, brines, grey water, black water, recycled water) will be available and employed for various purposes by multiple users.

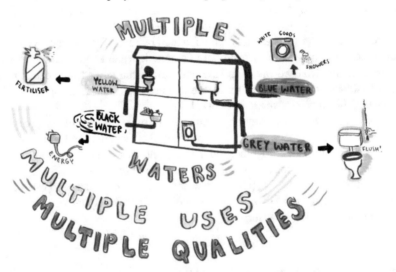

The water entering and exiting a domestic house serves as an illustration of this concept. Currently, water entering the dwelling is potable water supplied by the municipality or rainwater collected by the householder. The multiple waters generated from the house are shown in Fig. 1.2.

These consist of:

- **Greywater**, (i.e. used water from showers and wash hand basins, which do not contain faecal material)

- **Blackwater** (used water from toilets and the kitchen which contains faecal material)
- **Yellowwater** (used water from the toilets, urine separated from solid material by fitted devices).
- **Rainwater** - rainwater can be harvested and reused on site, either within the household for non-potable uses or for irrigation or to recharge local groundwater.

Fig. 1.2 shows a range of waters, referred to as multiple waters, available for multiple uses. For instance the greywater can be used for greywater recycling, replacing potable mains water for toilet use and reducing the demand for potable quality mains water. Yellow water may be used for plant or garden irrigation.

Figure 1.3 illustrates how the multiple water concept can be applied to a water supply network, resulting in a series of closed loops, where water from one process is reused in another, and where the quality of water governs its use.

Here we see multiple waters which can serve multiple uses. Rainwater falling on the roof or collected by the house holder can serve, with treatment, as an alternative source of potable water. Greywater, used water from showers and wash hand basins, which do not contain faecal material can replacing mains water for toilet use through the process of greywater recycling. Yellowwater, urine separated from solid material

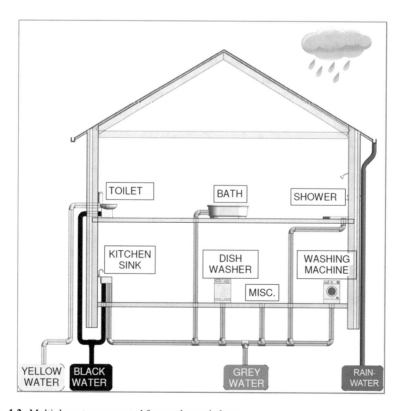

Fig. 1.2 Multiple waters generated from a domestic house

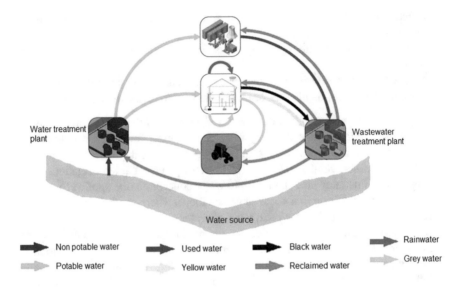

Non potable water	Used water	Black water	Rainwater
Potable water	Yellow water	Reclaimed water	Grey water

Fig. 1.3 The circular economy of water

by fitted devices, may be used for plant or garden irrigation. Black water, used water containing faecal material, can be used as fertiliser for non food chain plant products.

The CEW views industrial waters in the same way. Effluents from one industry contain resources that serve as influents for other industrial processes. Water containing dissolved organics and nutrients, serve as irrigation water for plant and agricultural production. Minerals which previously were removed by treatment can in this process serve as raw material for another industry resulting in the recycling of finite raw materials. In the linear economy these minerals would have been removed to landfill. Thus the CEW serves to recycle and reduce.

Another point of departure between the linear economy of water and the CEW is the quality of the water. In the linear economy all water was treated to potable water standards regardless of its use/function. Therefore water used for personal hygiene and life support was of the same quality as water used for carriage in the sewerage system or water used for urban street cleaning. Wastewater treatment standards also demanded a high quality water before discharge to a receiving water. The issue of quality in the CEW is more complex but also more sustainable.

The concept of multiple waters for multiple uses is different to the use of treated potable water for all applications. Multiple waters means that waters of different qualities are available for various processes. These multiple waters are characterised by their content, (water, heat, dissolved organics, minerals, nutrients) and it is this content that has multiple uses. Therefore an effluent from a pharmaceutical process may contain organic materials suitable for food production. An effluent from a mining process may be a suitable influent to a metal fabrication process. In the CEW the issue of water quality is replaced by the term "fit for purpose".

The concept of "fit for purpose" is well accepted in industry. However it has little acceptance with regulatory bodies and indeed environmentalists. As applied to industry it means that the quality of the water being used is of sufficient quality for the task in hand. The implication is that used water is only treated when it is of a lower quality than the task in hand. In effect when the used water is not fit for the purpose for which it is intended to be utilised. This "fit for purpose" concept therefore gives rise to different quality waters, which have different applications or uses.

This is not the uniform potable water regulatory regime of the twentieth century. It will require legislation to address the quality standards for used water. This is already the case in the European Union where draft regulations on Water Reuse are currently in the final stages of the legislative process. These regulations (as shown in Table 1.1) aim to separate reclaimed water into four different categories for usage. Waters are classified as Class A, food crops with edible part in contact with the water, Class B food crops with no contact with the reclaimed water, Class C

Table 1.1 Reclaimed water quality criteria for agricultural irrigation

Reclaimed water quality		Quality criteria				
Indicative technology target		E. *coli* (cfu/100)	BOD$_5$ (mg/l)	TSS (mg/l)	Turbidity (NTU)	
Class A	Secondary treatment, filtration, and disinfection (advanced water treatments)	≤10 or below detection limit	≤10	≤10	≤5	*Legionella* spp.: ≤1000 cfu/l when there is risk of aerosolization. Intestinal nematodes (helminth eggs): ≤1 egg/l when irrigation of pastures or fodder for livestock
Class B	Secondary treatment, and disinfection	≤100	According to Directive 91/271/ EEC	According to Directive 91/271/ EEC	–	
Class C	Secondary treatment, and disinfection	≤1000	According to Directive 91/271/ EEC	According to Directive 91/271/ EEC	–	
Class D	Secondary treatment, and disinfection	≤10,000	According to Directive 91/271/ EEC	According to Directive 91/271/ EEC	–	

food crops with drip irrigation and Class D where the reclaimed water is used industry, energy or seeded crops. All waters regardless of class require the use of secondary treatment and disinfection, with Class A reclaimed water requiring additional filtration.

The aim of the legislation is to increase the amount of wastewater reused in Europe. At present, about 1 billion cubic metres of treated urban wastewater is reused annually, which accounts for approximately 2.4% of treated urban wastewater. The EU reuse potential is much higher, estimated at up to 6 billion cubic metres per year. This has potential to provide a reliable water supply, independent from seasonal droughts. The importance of water standards for reclaimed water use will also be important in promoting the acceptance by the public at large of the use of reclaimed water and eventually multiple waters. Some countries and some cultures may find it initially difficult to accept produce used from

ONLY 10% OF WATER USED IS DRINKING WATER... SO WHY IS IT ALL DRINKING STANDARD?

reclaimed water. Water standards are a way of safeguarding the consumer in countries that are concerned about the safety of reclaimed water use. Other significant issues will focus on the accountability for the quality of reclaimed water along the distribution chain. Here the end of responsibility for the reclamation plant operator will be after the point of compliance. However the problems arising out of climate change, with periods of increasing water scarcity together with greater demand on existing water resources will probably have more of an impact on the public acceptance of multiple waters than will standards for reclaimed water.

1.5 The Worth of Water

Water is more than just a commodity; it is a basic human right. It is in fact a life-support medium. We cannot live without it. Industry cannot function without water and therefore, economically, water is necessary for society to exist. High-quality water is necessary for health. Water is also required for energy and is an important amenity for recreation and indeed rural income. These functions of life support, industrial use, energy supply and amenity can be termed the **"Value of Water"**.

The **"Value in Water"** can be thought of as the economic and societal value that can be realised by extracting and valorising the resources embedded in used water streams. These include the nutrients, minerals, chemicals, metals and energy which are embedded in water. It also refers to the reuse potential of the water. There are many examples of extracting the embedded value in water. Ardagh Group, Ireland at its Dongen plant in the Netherlands recycles calcite, a waste product from the water treatment industry to replace limestone as a raw material in the production of high quality container glass. In Denmark, the Billund Bio Refinery is another example of the circular economy of water. Wastewater is treated to produce clean water, with the sludge recovered and treated to become high-value products like organic fertiliser, phosphorus, bioplastics and also biogas, which goes on to produce clean electricity and heat.

So moving from a narrow discussion of the price of water, we start to explore the value in water and this deeper discussion leads us to consider the **"Worth of Water"**. A dictionary definition of worth describes it as the *importance* or *usefulness* of something. The importance of water is that it is a life-support medium; we cannot live without it, and this is the value *of* water. The usefulness is the potential for resource recovery and reuse, or the value *in* water. We can therefore define the worth of water as the value of water combined with the value in water.

How do we deal with both the value of water and the value in water? What measures do we require to take to move from the existing approach to water to adopting an approach that fully appreciates the worth of water?

Chapter 2
Properties of Water

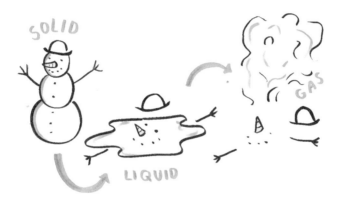

2.1 Introduction

Water is a unique compound. It surrounds us, gives us life and without it civilisation would cease to exist. When we consciously encounter it first, normally in our early youth, it is a source of wonder. However, the familiarity we gain from drinking it, washing with it, cooking with it, soon removes that initial sense of wonder, and we tend to take water for granted. It is only when water impinges on our life that we notice its characteristics, and how unique these characteristics are. These characteristics are known as the properties of water and they can affect our daily life.

Those of us who went to school in temperate regions have experience of the expansion of water on freezing which is a problem when water is confined in pipes or tanks, and no allowance is made for this expansion. The high heat energy required to warm water has played a role in moderating the Earth's climate and the effect that increased temperatures have on the global system. The high surface tension of water

© Springer Nature Switzerland AG 2021
L. McCarton et al., *The Worth of Water*,
https://doi.org/10.1007/978-3-030-50605-6_2

means we require a detergent when washing a greasy plate. Water dissolves most substances (but not all) and is therefore known as the universal solvent.

All these properties and more, are explained by the molecular structure of water.

2.2   Molecular Structure of Water

Water is formed when an atom of oxygen combines with two atoms of hydrogen. The result is a substance that is very different from most other substances. This is due to the way the atoms combine and arrange themselves. An atom is a neutral entity, possessing neither a positive nor negative charge. Molecules are also neutral, though when they lose or gain electrons (negatively charged atomic particles), they can become charged and are then referred to as ions.

With hydrogen and oxygen, their combination is influenced by their valency, which is their combining power. Elements combine in order to fill their outer shell. They do this by sharing or exchanging electrons. The valency refers to the number of electrons each element requires to lose, gain or share to complete its outer shell. Hydrogen requires one electron in order to fill its outer shell to two. Oxygen, which has six electrons in its outer shell, requires two, in order to fill that shell to eight. Therefore, the one electron required by hydrogen is supplied by oxygen, and the two required by oxygen is supplied by two hydrogen atoms, resulting in the molecular formula for water of H_2O. Figure 2.1 illustrates the molecular structure of water.

This sharing of electrons is known as a covalent bond. A covalent chemical bond consists of a pair of electrons shared between two atoms. Thus, the oxygen atom is surrounded by four electron pairs that would ordinarily tend to arrange themselves as far from each other as possible. However, this does not happen. As a result of

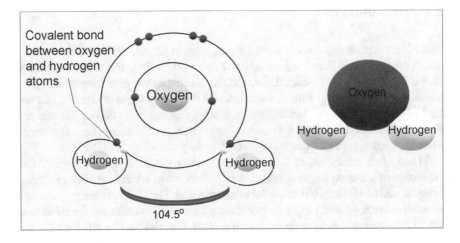

Fig. 2.1 Molecular structure of water

Electronegativity, which is the tendency of an atom to attract a bonding pair of electrons, the electrons in the water molecule are not shared equally between hydrogen and oxygen. The difference in electronegativities of oxygen and hydrogen is 1.4, resulting in the shared electrons being drawn preferentially towards the oxygen atom.

The water molecule is electrically neutral, but the positive and negative charges are not distributed uniformly. The highly electronegative oxygen atom attracts electrons or negative charge to it, making the region around the oxygen more negative than the areas around the two hydrogen atoms. This charge displacement constitutes an electric dipole, and it makes water about 30% ionic in character. As a result, the water molecule is considered positively charged at hydrogen ends of the molecule and negatively charged at the oxygen end. Opposite charges attract, and this is what gives rise to another characteristic of the water molecule. Figure 2.2 shows the polarity of the water molecule.

The polarity of the water molecule results in a partially positive hydrogen atom on one water molecule being attracted to a partially negative oxygen on a neighbouring molecule. This is called **hydrogen bonding** and is illustrated conceptually in Fig. 2.3. The hydrogen bond is weak and can only survive for more than a

Fig. 2.2 Polarity of water molecule

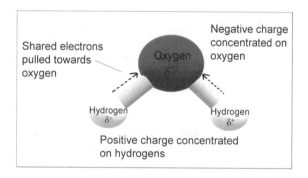

Shared electrons pulled towards oxygen

Negative charge concentrated on oxygen

Oxygen

Hydrogen

Hydrogen

Positive charge concentrated on hydrogens

Fig. 2.3 Hydrogen bonding does not last very long

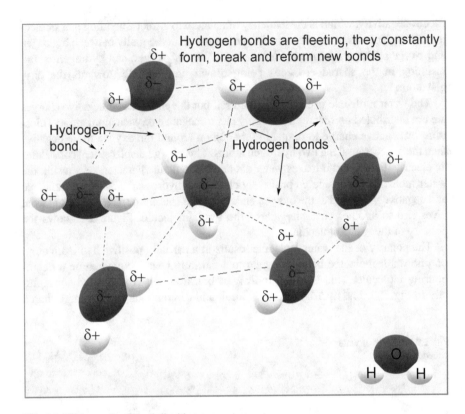

Fig. 2.4 Hydrogen bonding in liquid water

tiny fraction of a second. However, there are so many of them forming and breaking at any one time that they play a very important part in the properties of water. Figure 2.4 shows hydrogen bonding in liquid water.

The unique shape of water is due to the two lone pairs of electrons, supplied by Oxygen, in the water molecule, H_2O. The water molecule has a bent shape, which is explained by the two lone pair of electrons being on the same side of the molecule. They repel each other, causing the bond of the hydrogen to the oxygen to be pushed downward (or upward, depending on your point of view).

2.3   The Properties of Water

Water has long been known to exhibit many physical and chemical properties that distinguish it from other small molecules of comparable mass. Chemists refer to these as the "anomalous" properties of water, but they are by no means mysterious. All are entirely predictable consequences of the ***shape of the water molecule, the polarity of the water molecule and hydrogen bonding.***

2.4 The Three States of Water

The only substance that exists, in three states, as a solid liquid or gas, within the normal temperature range found on the Earth's surface is water. The solid phase, ice, is characterised by hard packed crystals, ice, or loose crystals, snow. The liquid phase is what we refer to as water. The gas is known as water vapour or steam and is generally a transparent cloud.

Figure 2.5 illustrates the three states of water. Ice is converted to water by melting. Water is converted to steam by evaporation. The reverse process is also shown where condensation of steam results in liquid water and freezing of water results in the formation of ice. The conversion from steam directly to ice is known as deposition and from ice directly to steam as sublimation.

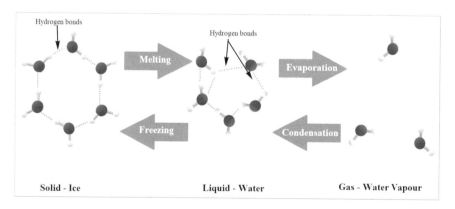

Fig. 2.5 The three states of water

2.5   The Global Water Cycle

the WATER CYCLE This ability to exist as ice, water or water vapour is the basis of the water cycle. It is by means of the water cycle that water is circulated from the ocean to the land and back to the ocean. The ability of water to change states is the critical factor here, as water is evaporated from the surface of the earth and rises to the atmosphere as water vapour. In the atmosphere it forms clouds and falls to the earth as precipitation. At the poles and at high altitudes this precipitation falls as snow. This allows for a renewable supply of fresh water. It also drives the global and local weather patterns and rainfall. Figure 2.6 illustrates the water cycle.

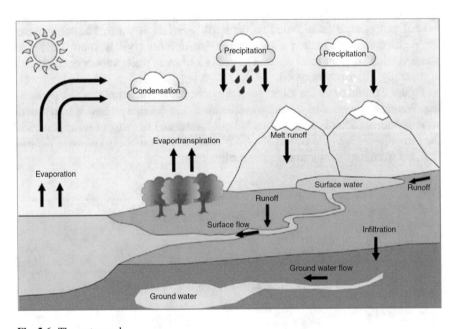

Fig. 2.6 The water cycle

Global Water Cycle

The amount of water on the Earth is a fixed quantity. The planet is home to 1.386 million cubic kilometres of water. However, of this volume, a very small quantity, or 2.5% is fresh water. The other 97.5% is salt water. Of this small fraction of water that is considered fresh, 35 million cubic kilometres of it, less than a third is technically available to humankind. What is regarded as technically unavailable fresh water, equivalent to 69.5% of total fresh water, is tied up in glaciers, snow, ice and permafrost. This leaves almost 11 million cubic kilometres of freshwater available to us humans for our activities. Taking the world's population as 7.8 billion people, enough rain falls to provide, on average, 7000 m^3 of fresh water per annum. This is more than enough for most needs, however the rain does not fall equally nor are people free to move to areas of abundant water. Therefore, areas and populations have water scarcity problems. Only a third of the world's population has a sufficient or plentiful supply of water. Another third of the population have insufficient water, while a quarter of the world are in a situation of water stress, where water supplies are insufficient to meet daily demands. Almost 10% of the world's population have a scarcity of water.

2.6 The Density of Water

Density refers to the amount of matter in a given volume of a substance. It generally increases on cooling a liquid while it decreases with increasing temperature. Water, like other substances contracts when cooled until it reaches 4 °C. However, when water molecules freeze together they link up in chains that are not as compact as the chains in liquid water. In a water molecule the distance between each Hydrogen (H) and Oxygen (O) atom is 0.099 nanometre (nm). When ice forms, hydrogen bonds some 0.177 nm long develop between the H and O atoms in adjacent molecules. This produces a crystal lattice structure which has a greater volume than a liquid water molecule, and hence is less dense. This reduces the density of the ice to less than that of the water. Its lower density means that ice floats on water.

The importance of this phenomenon to life cannot be overstated. This is the reason why lakes only freeze at the surface. The ice forms an insulating layer at the surface of ponds and lakes preventing any deep bodies of water from freezing solid top to bottom, preserving fish and plants below the ice. The ice easily melts in the spring supporting new aquatic life. It also allows animals and humans to survive on the fish life in the water bodies.

The expansion of water on freezing is a problem when water is confined in pipes or tanks, and no allowance is made for this expansion. In colder climates this can often result in pipes or containers bursting, a common winter bonus for school children. The phenomenon of spring and autumn turnover in lakes is also explained by the change in water density due to temperature.

This is a very important phenomenon to preserving ecosystems. Ice forms an insulating layer on the surface of ponds and lakes preventing any deep bodies of water from freezing solid top to bottom. This protects fish and plants below the ice. The ice easily melts in the spring supporting new aquatic life. Figure 2.7 illustrates the crystalline structure of ice showing some hydrogen bonding.

The hue of the water in the photo is due to the particles in the water produced by the grinding action of the glacier on the mountain not settling. These colloidal particles give mountain lakes and glacier fed lakes this distinctive colour (Fig. 2.8).

Where two different substances are mixed with one less dense than the other one, the less dense substance floats on top of the more dense substance. Hence, ice floats in a glass of water, and acts to cool it.

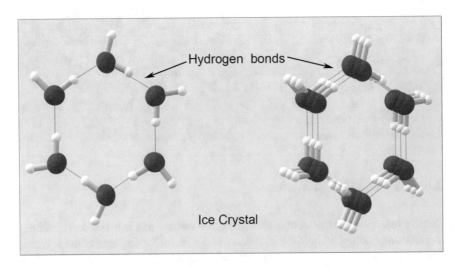

Fig. 2.7 Crystalline structure of ice showing some hydrogen bonding

Fig. 2.8 The distinctive colour is due to colloidal particles

2.7 Water and Heat

The uniqueness of water is again displayed in its reaction to heat. The experience of milk boiling over is common with those who live in temperate climates. This can be explained by the term specific heat capacity and by the fact that water has a greater specific heat capacity than milk. The term specific heat capacity refers to the amount of heat required to raise a kilogram of a substance by 1 °C. Water has a very high specific heat. This means it takes more heat to raise its temperature, than most substances encountered in daily life. Water also has an abnormally high heat of vaporisation, which refers to the heat required to convert it from the liquid to the gaseous state, water to steam.

These two properties also play an important role in the influence global warming is having on the planet. The high heat energy required to warm water plays a role in moderating the Earth's climate and the effect that increased temperatures have on the global system. The oceans have absorbed a lot of the excess heat due to global warming and this has buffered the global system against large fluctuations in temperature. The oceans can absorb approximately a thousand times more heat than the atmosphere without changing temperature.

Global Warming and the Oceans
Recent studies estimate that warming of the upper oceans accounts for about 63 per cent of the total increase in the amount of stored heat in the climate system since 1971. The full depth ocean heat gain rate ranges from 0.57 to 0.81 watts per square meter. When this is multiplied by the surface area of the ocean (360 million square kilometres) this translates into an enormous global energy imbalance. In the present, warming of ocean water is raising global sea level because water expands when it warms. Combined with water from melting glaciers on land, the rising sea threatens natural ecosystems and human structures near coastlines around the world. Warming ocean waters are also implicated in the thinning of ice shelves and sea ice, both of which have further consequences for Earth's climate system. Finally, warming ocean waters threaten marine ecosystems and human livelihoods. For example, warm waters jeopardize the health of corals, and in turn, the communities of marine life that depend upon them for shelter and food. Ultimately, people who depend upon marine fisheries for food and jobs may face negative impacts from the warming ocean.

Source: www.climate.gov

The heat of vaporisation of water also plays an important role in regulating body heat. When the ambient temperature is above body temperature, heat is transferred into the body rather than out. Since there must be a net outward heat transfer, the only mechanisms left under those conditions are the evaporation of perspiration from the skin and the evaporative cooling from exhaled moisture. Even when one is unaware of perspiration there is a loss of moisture from the skin.

Water also has a high latent heat of fusion. This refers to the quantity of heat required to melt 1 kg of ice and of all the common substances, only that of ammonia is higher. This property slows the melting of ice and imparts resistance to melting of ice on glaciers. As a result of the high latent heat of fusion of ice, and the fact that all the ice must be converted to liquid water before temperature increase can occur, it is still used to chill drinks.

2.8 Cohesion and Adhesion

Cohesion is described as "like" molecules sticking together. Adhesion is where unlike molecules stick together. The polar nature of water and its resulting hydrogen bonding plays a part in these properties. Water molecules stick together due to cohesive forces and stick to other substances due to adhesive forces. Though hydrogen bonds are constantly forming, breaking and reforming, there are enough bonds at any one time to allow for cohesion and adhesion to occur. These forces are illustrated further in Figs 2.9 and 2.10.

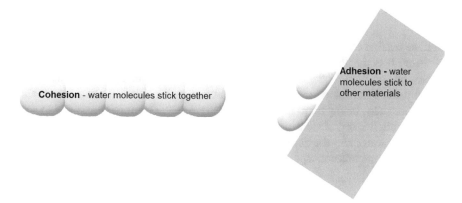

Fig. 2.9 Cohesion and adhesion

Fig. 2.10 Cohesion causes
water to form droplets,
adhesion allows water
droplets to stick to leaves

2.9 Surface Tension

Surface tension can be thought of as a skin formed by the molecules at the surface of water. It is the ability of a liquid to resist an external force. Surface tension is caused by cohesion. Water molecules at the liquid surface have no water molecules above them. They experience a stronger force exerted on them from adjacent and inner molecules below them in the body of the liquid. This force gives rise to surface tension. It is surface tension that allows water drops to keep their shape, insects to walk on water and the ability to float a needle on water. Figures 2.11 and 2.12 further illustrate this concept.

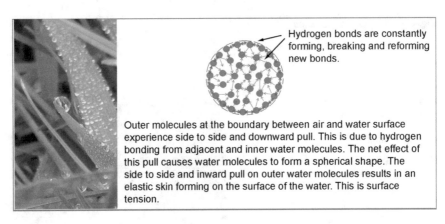

Hydrogen bonds are constantly forming, breaking and reforming new bonds.

Outer molecules at the boundary between air and water surface experience side to side and downward pull. This is due to hydrogen bonding from adjacent and inner water molecules. The net effect of this pull causes water molecules to form a spherical shape. The side to side and inward pull on outer water molecules results in an elastic skin forming on the surface of the water. This is surface tension.

Fig. 2.11 Surface tension

Fig. 2.12 Example of surface tension

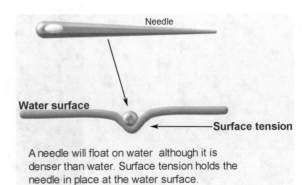

Needle

Water surface

Surface tension

A needle will float on water although it is denser than water. Surface tension holds the needle in place at the water surface.

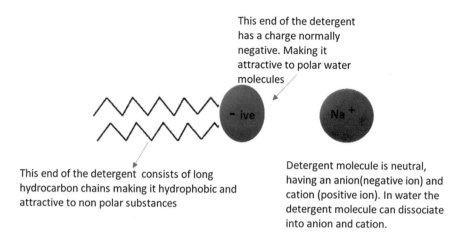

This end of the detergent has a charge normally negative. Making it attractive to polar water molecules

This end of the detergent consists of long hydrocarbon chains making it hydrophobic and attractive to non polar substances

Detergent molecule is neutral, having an anion(negative ion) and cation (positive ion). In water the detergent molecule can dissociate into anion and cation.

Fig. 2.13 Detergent molecule

When using water for cleaning we must counteract surface tension. Take the example of washing a greasy plate. When water is poured on to the plate, the grease being non-polar repels the water, it forms in droplets on the grease, neither wetting the surface nor spreading on it. A detergent is required to reduce/remove the surface tension of the water, in effect opening the skin formed by the surface tension. A detergent is a molecule with two distinctive ends, one hydrophilic (water loving) and the other hydrophobic (water hating). The hydrophobic end attaches to the grease, while the hydrophilic end attaches to the water molecules allowing the water to wash away the grease. Figure 2.13 shows this concept.

As a result of the forces of adhesion and surface tension, water can rise up a narrow tube, against the force of gravity. This is known as capillary action and occurs due to water adhering to the inside wall of the tube and surface tension straightening the surface causing a surface rise and more water is pulled up through cohesion. The process continues as the water flows up the tube until there is enough water such that gravity balances the adhesive force.

Surface tension and capillary action along with cohesion and adhesion play important roles in biology. For example, when water is carried up stems in plants, through the xylem, the strong intermolecular attractions (cohesion) hold the water column together and adhesive properties maintain the water attachment to the xylem and prevent tension rupture caused by pull from transpiration, i.e. where plants absorb water through the roots and then give off water vapour through pores in their leaves. Figures 2.14 and 2.15 illustrate this further.

Fig. 2.14 Capillary action

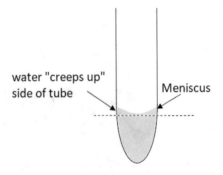

water "creeps up"
side of tube   Meniscus

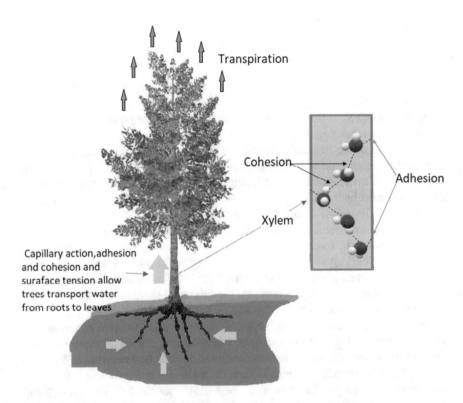

Transpiration

Cohesion

Adhesion

Xylem

Capillary action,adhesion
and cohesion and
suraface tension allow
trees transport water
from roots to leaves

Fig. 2.15 Water transport in plants

2.10   Water as a Universal Solvent

A substance which is dissolved is called a solute. The substance which does the dissolving is called a solvent. The result is a solution.

The polar nature of water means that it is a very good solvent. Substances can be divided onto those that mix well in water and are called hydrophilic (water loving) and those that do not mix with water called hydrophobic (water repelling) substances. To dissolve in water a substance must generate attractive forces that are stronger than the attractive forces between water molecules. Where a substance does not attract a water molecule with greater force than that within the water molecule itself, the molecules are pushed together by the water and the molecules of the substance do not dissolve.

A hydrophobic substance is one that is non-polar whose molecules stick together to exclude water molecules in an aqueous solution. Figure 2.16 shows an example of this using oil and water.

Hydrophobic reactions are critical at cellular level and are the foundation of the existence of life.

Hydrophilic molecules form ionic bonds or hydrogen bonds with water molecules. This occurs as the small size of the water molecule allows many molecules of water to surround one molecule of solute.

The partially negative ends of the polar water molecule are attracted to the positively charged components of the solute, and vice versa for the positive ends of the polar water molecule.

It is because of this polarity, this ability to surround positive ions with the negative end of the water molecule and to surround negative ions with the positive end of the polar water molecule, that water is referred to as the universal solvent. Figure 2.17 shows this concept.

Water molecules are polar. Olive oil molecules are non-polar i.e. they have no charge separation

The water molecules are more attracted to each other than to the oil molecules.

The oil molecules are attracted to each other rather than water molecules.

Olive oil with some water added

Water and oil do not mix. They are immiscible liquids.

As olive oil is less dense than water, it floats on water.

Olive oil

Water

Fig. 2.16   Oil and water – immiscible liquids

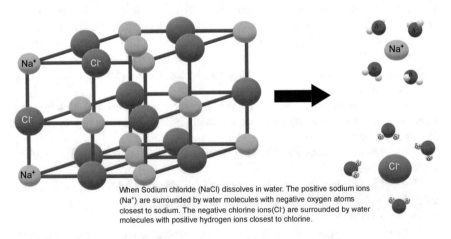

When Sodium chloride (NaCl) dissolves in water. The positive sodium ions (Na$^+$) are surrounded by water molecules with negative oxygen atoms closest to sodium. The negative chlorine ions(Cl$^-$) are surrounded by water molecules with positive hydrogen ions closest to chlorine.

Fig. 2.17   Water as a universal solvent

Chapter 3
Removal Mechanisms

3.1 Introduction

Potable water refers to water that is fit to drink. It is fit for human consumption and to classify as such it must meet legal standards characterised by water quality parameters. These parametric values are typically met by unit processes which form part of potable water treatment. Here physical, chemical and biological removal mechanisms are utilised to remove pollutants. These removal mechanisms are not confined to use in water treatment. They are also utilised in wastewater treatment and rainwater harvesting. Conventional roof rainwater harvesting systems utilise these same removal mechanisms in the treatment train that exists from the capture of the water on the roof through the filter, the pump and the storage tank. Thus, the removal mechanisms utilised in potable water treatment are mirrored in the treatment train of rainwater harvesting systems.

© Springer Nature Switzerland AG 2021
L. McCarton et al., *The Worth of Water*,
https://doi.org/10.1007/978-3-030-50605-6_3

3.2 The Water Treatment Process

The aim of water treatment is the production of water that is safe, clear and aesthetically pleasing. Safe is taken to mean free of pathogenic or disease-causing organisms while aesthetically pleasing is usually taken to mean free of odours and tastes. Water treatment plants treat raw water by a sequential series of unit processes aimed at removing unsafe or displeasing parameters in the water. A water treatment plant is shown in Fig. 3.1 below.

Raw water is normally stored either before treatment or immediately after screening. Storage serves three important roles. Firstly it allows for the balancing of supply and demand Secondly, storage can facilitate the settling of suspended solids in the raw water and thirdly exposure to sunlight gives rise to ultra-violet (UV) treatment which can reduce pathogenic content and also reduce any colour in the water. Typically the first part of the treatment is screening of the incoming raw water. This is aimed at removing gross solids and any suspended solids carried in the water. Gross solids refers to larger solids that are not considered part of the water and generally refers to trees, branches, animal carcasses, oil slicks and other constituents that are not part of the water. Suspended solids refer to all settleable solids in the water. These latter interfere with the passage of light through the water and may also be pathogenic. The size of screen is usually governed by local conditions such as rainfall and solids loading. The next stage of treatment is the removal of colloidal material. Colloids are solutions that contain particles that will not settle. This is due to all particles possessing the same charge, either all positive or all negative charges. As a result the particles repel each other and even after stirring will not settle as they move as far apart from each other as possible. The process of colloidal removal is a two-step process consisting of coagulation and flocculation. Coagulation is the addition of a chemical, commonly aluminium

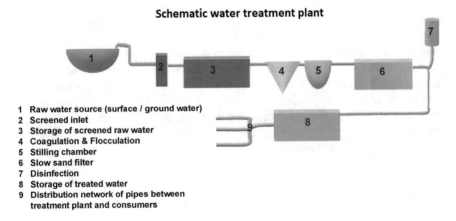

Schematic water treatment plant

1 Raw water source (surface / ground water)
2 Screened inlet
3 Storage of screened raw water
4 Coagulation & Flocculation
5 Stilling chamber
6 Slow sand filter
7 Disinfection
8 Storage of treated water
9 Distribution network of pipes between
   treatment plant and consumers

Fig. 3.1 Typical removal mechanisms within a water treatment plant

sulphate or ferric chloride. This removes the charges from the particles. Flocculation (this is a process where the smaller particles come together to form larger particles which can then be removed by settlement or filtration) then comes into play as the particles, now possessing no charge begin to gather and a floc starts to form. The water then passes to a stilling chamber where the flocculant is settled and removed. The treated water is then passed through a slow sand filter where it is clarified by particle removal. Disinfection is generally the last process performed and this usually entails the addition of chlorine or the use of ultraviolet radiation, reverse osmosis or other disinfection process. The parameters removed, the physical, chemical and biological removal mechanisms and the technologies employed are shown in Table 3.1.

Table 3.1 Physical, chemical and biological removal mechanisms utilised in a typical water treatment plant and some of the technologies employed in removing the various parameters

Treatment	Function	Parameter removed	Removal mechanism	Water treatment plant	Rainwater harvesting non potable supply	Rainwater harvesting potable supply
Preliminary	To protect the Water entering system	Suspended solids	Screening	✓	✓	✓
Storage & aeration	Balance and Supply	Suspended solids, pathogens, colour	UV, sedimentation, settlement, bleaching	✓	✓	✓
Coagulation, flocculation	Addition of chemicals to remove colloidal particles	Colloids	Charge removal and particle adherence, precipitation	✓	☐	☐
Sedimentation & filtration	Removal of suspended solids and clarification	Suspended solids, colour, turbidity, pathogens	Sedimentation and filtration	✓	✓	✓
pH adjustment	To alter the acidity / alkalinity	Hydrogen ion or Hydroxyl ion	Chemical addition, precipitation	✓	☐	☐
Disinfection	Produce safe potable water	Pathogens	Ozone, ultraviolet, micro-filtration, chlorination	✓	☐	✓

The final columns show the removal processes involved in most rainwater harvesting installations, be they for potable or non-potable use.

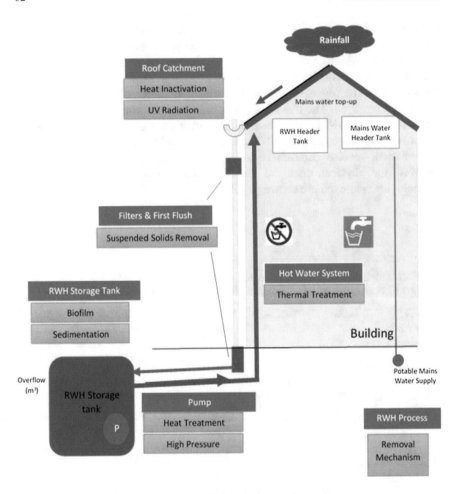

Fig. 3.2 Removal mechanisms within a typical roof rainwater harvesting system

3.3   The Rainwater Harvesting Treatment Train

Figure 3.2 illustrates the removal mechanisms which are integral to a typical domestic roof rwh installation. It shows the locations at which the water treatment takes place and the particular removal mechanisms involved. This treatment is performed by the removal mechanisms which remove various parameters in the rainwater as it proceeds through the rwh system. Removal mechanisms taking place in the tank are treated separately below.

These removal mechanisms are responsible for what is known as an inherent water treatment process or 'treatment train' in the rainwater harvesting process.

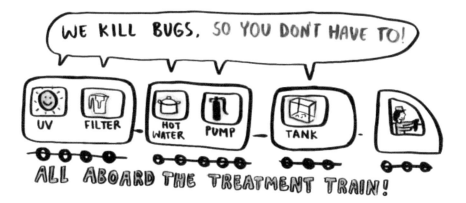

3.3.1   Roof Catchment

The roof is the first site of the treatment train. The removal mechanisms involved are heat inactivation and ultraviolet radiation. The higher the temperature the greater the effect on bacterial removal. UV radiation due to sunlight also plays a part in the reduction of the microbial population. Tropical countries with intensive sunlight and extensive hours of sunshine will see increased rates of microbial removal when compared with more temperate zones. Set against the removal effect of temperature and UV radiation is the fact that the roof can serve as a collection centre for various solids (dust particles, windblown debris etc) and microbiological substances ranging from bird and rodent droppings to leaf and organic matter decay. However, a major proportion of this input is removed in the filtration device.

3.3.2   First Flush and Filters

The function of a filter device is to remove any solids material carried off the roof by the rainwater. The removal mechanism is filtration. The filter device may be a commercially available stainless steel construct, of a set filter aperture or it may be a more rustic device referred to as a first flush device. Studies have shown that both these devices, can remove 11–94% of dissolved solids and 62–97% of suspended solids from the first 0.25 mm of roof runoff.

3.3.3 Pump

The removal mechanism here is heat treatment. Where high pressure pumps are used removal of substances may occur due to high pressure.

3.3.4 Hot Water System/Cistern

Hot water supplies are normally required to reach a minimum of 60 °C. This was introduced to combat legionellosis. Therefore, when a water is passed through a hot water system/cistern it is subject to temperatures in excess of those required for pathogenic survival. Thus the removal mechanism operating in the hot water system is thermal inactivation of bacteria and pathogens.

3.3.5 The Storage Tank

Rainwater storage tanks function as ecosystems with self-regulating processes. These removal mechanisms include settlement, flocculation, competitive exclusions and the growth of biofilms. As shown in Fig. 3.3 a rainwater stored in a tank shows different behaviours at various locations in the tank. The heavier particles are removed by settlement. Flocculation removes organic, metallic and chemical pollutants. In an ecosystem where few nutrients are available there is competitive exclusion of bacteria. At these low nutrient concentrations, there are advantages for microbes to form attachments to containment surfaces. Therefore biofilms form in rainwater tanks. They also develop a complexity and this allows for significant removal of bacteria and metals from the water column. Studies show that the bacteria numbers in the biofilm were 1280 to 14,060 times the bacteria count in the water column, where bacteria are removed from the water.

Biofilms consist of a core of environmental bacteria such as *Bacillus Sp*. These studies also showed that concentrations of lead in the biofilm was 250–125,000 times the concentration of lead in the water column. The sludge settled at the bottom is also a biofilm. Typically studies showed deposition rates of 0.6–7.8 mm/year with *E.coli* in the sludge 21-1350 times that in the water column with lead values in the sludge being 66,680–343,000 times the water column. Therefore the storage tank plays a notable role in the rainwater harvesting treatment train. It also shows the usefulness of not removing sludge deposits.

Fig. 3.3 Rainwater
storage tank functions as a
bio-reactor

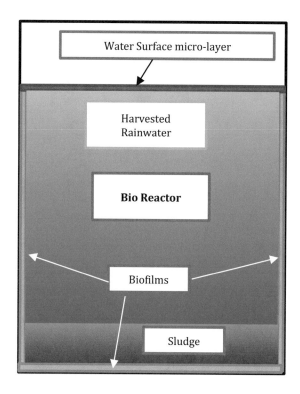

Chapter 4
The Worth of Water Technology Portfolio

4.1 Introduction

The importance of water is that humans cannot live without it. Insufficient water results in the human body collapsing and dying. This is considered the value of water. This value also extends to commerce and agriculture. The value in water is a concept related to the circular economy of water and refers to the resources in used water. These resources include nutrients, heat, minerals and of course water itself. Taken together, the value of water and the value in water give us the worth of water. The social life of water is a term recently proposed to encompass all domains of human relationships with water. This includes the political relationship, economics, the technological interaction with water and the spiritual. As with the concept of the

© Springer Nature Switzerland AG 2021
L. McCarton et al., *The Worth of Water*,
https://doi.org/10.1007/978-3-030-50605-6_4

worth of water the social life of water views water and humanity's dependency on water as complex and involving all aspects of everyday human existence.

These may be new concepts introduced to help us gain a holistic understanding of the importance of water to modern day society. However, ancient civilisations and their society recognised the importance of this interconnectedness of the natural world and the human world. Cultures ranging from Iran, to Peru, Sri Lanka to the Yucatan understood the need to live in balance with natural water resources. They understood intuitively the importance of water to the continued existence of their cultures. Where this delicate balance was broken, societies entered a system of collapse.

There is a stark contrast between these cultures and where the world is with regard to water today. Almost a billion people live in shantytowns with no access to safe drinking water and no adequate sanitation. An increasing world population requires intensive water dependent agriculture to feed 9 billion people by 2050. The worth of water is in danger of becoming commodified with its availability being confined to the rich of the world. Further issues of water quantity, water quality and access to water stand in stark contrast to cultures that are presented in this chapter. This chapter presents a technology portfolio of case studies which all illustrate the worth of water.

4.2   Qanat Civilisation

The Worth of Water Case Study: Qanat Civilisation
Source: Majid Labbaf Khaneiki, International Centre on Qanats and Historic Hydraulic Structures, www.icqhs.org

A qanat is an ancient technique developed in arid or semi-arid regions to extract groundwater for irrigation and domestic use. It is believed that qanats were developed originally in Iran to enable communities to survive in the harsh arid conditions. The technology quickly spread to neighbouring countries. Today qanats remain an important water supply infrastructure in countries such as Iran, Afghanistan, Azerbaijan, Iraq, Oman and Pakistan. Qanat construction technology was handed down from father to son and resulted in the accumulation of specialised knowledge on water resource management and lead to the build up of a distinct civilization – Qanat civilisation.

A qanat comprises of a horizontal tunnel with a gentle slope that partly cuts through the aquifer. These tunnels can continue for several tens of kilometres, with minimal water loss from evaporation or leakage. Water within the saturated layers seeps into the tunnel and flows according to the gentle gradient. Vertical shafts are excavated along the length of the qanat to provide a means of ventillation, as well as a way to bring debris and excavated material to the surface. Water drains by gravity and does not require any mechanized pumping. The qanat system drains the overflow of groundwater from the aquifer and therefore does not affect the water balance unlike deep wells and boreholes. Water extraction by means of the qanat does not result in any lowering of the water table or consolidation of the soil. The qanat system supplies water for irrigation and via a network of water reservoirs for drinking and other domestic use. Figure 4.1 shows a cross section through a typical qanat system.

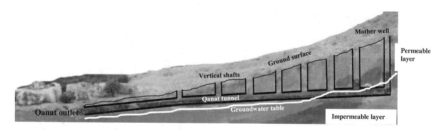

Fig. 4.1 Cross section through a qanat showing the horizontal qanat tunnel tapping the groundwater aquifer above the impermeable layer, mother well, vertical shafts and qanat outlet to distribute water to users

A qanat construction team typically comprised of between 2 and 6 people and had a high social position and a place of respect within society as valued water professionals. As a result, even though a dangerous profession, there was no shortage of qanat teams with many handing down the profession from father to son. The qanat team used a pickaxe, pulley, rope, shovel, carbide lamp, compass and plumb line to dig the qanat and a windlass to haul the debris up the tunnel. Figure 4.2 shows photos from the qanat museum in Yazd to illustrate the techniques.

Fig. 4.2 Photos from the Qanat Museum in Yazd, Iran showing qanat teams and their equipment excavating the vertical shafts for a qanat

Qanats and Sustainability

Water resource management using the system of qanats ensured that a sustainable balance was maintained between water supply and demand. The qanat did not over exploit the available water resources and did not result in lowering of the ground water table. In periods of extreme drought, the qanat users had two options. To increase supply they could extend the qanat tunnel back through the aquifer. Alternatively they could reduce the water demand area of the farm. Each farm was divided into two areas, one for trees and the other for crops. In periods of droughts, they adjusted this cultivation ratio to the available water supply in order to ensure a balance was maintained.

Qanat "Water Shares"

In most cases, ownership of the qanat is separate from land ownership. Each qanat may belong to hundreds of farmers who are assigned a number of *"water shares"* which enables them to take turns getting their water to irrigate. Shareholders were allocated a share (or water volume) according to a special type of water clock called a *"clepsydra"*.

This consisted of two bowls made of copper, one small inside one larger. The smaller one floats on the surface with a tiny hole in the base. Water gradually fills the smaller bowl and it sinks to the bottom with a bump. The unit of "water time" was the time

Fig. 4.3 Traditional qanat water clock "clepsydra"

between two bumps. A complex structure developed depending on the rotation of irrigation rights, the season, time of flow in the qanat and water shares. For example, a shareholder with 10 shares, each worth 4.6 minutes and a rotation of 21 days, would be entitled to irrigate their land for 46 minutes every 21 days. Figure 4.3 shows a traditional qanat water clock.

Water Distribution Structure (Maqsam)

At the outlet of the qanat there is a structure called a maqsam comprising different size weirs to distribute the water according to the shareholders agreement. Due to the complexity and number of shareholders involved, a water professional, called a "mirab" is often paid by all shareholders to ensure that the system works fairly.

Qanats Today

Unfortunately the profession has seen a decline in recent years. Young people see it as a low skilled manual job, with high levels of danger and no insurance. To preserve the legacy of the Qanat civilisation the Iranian authorities established the International Center on Qanats and Historical Structures under the auspices of UNESCO (www.icqhs.org). A training centre for qanat technicians was established in Yazd, Iran in 2005 with a two year course in qanat maintenance and rehabilitation in an attempt to maintain and upkeep the qanat hydraulic infrastructure upon which so much of Iran depends.

Qanat of the Moon, Ardestan

The only two level qanat ever known is the Qanat of the Moon, Ardestan, Iran. This complex qanat is considered a masterpiece example of ancient hydraulic infrastructure. It is a double tunnel qanat with two tunnels running parallel. An impermeable clay layer spans the 3 m vertical distance between qanats. The length of the moon qanat is 2 km, the number of shafts is 30 and the discharge is 60 l/s. The cycle of irrigation of this qanat is once every 10 days. This cycle consists of 1320 water shares. Each day (24 hours) is equal to 132 shares. The water source of the upper qanat and lower qanat is not the same so their taste and colour is different. Figure 4.4 shows the payab of this qanat which has two levels. Figure 4.5 shows the qanat gallery in the upper qanat channel inside the payab. These two streams of water join each other after they reach the surface in Ardestan town. Figure 4.6 shows the water

Fig. 4.4 A payab is a sloping gallery, connecting the ground surface to the qanat's gallery

Fig. 4.5 Qanat gallery

Fig. 4.6 Water distribution structure at qanat outlet dividing water into two thirds: one third

distribution structure (maqsam) with the water now divided into three parts: two parts belong to the farmers in Moon in Ardestan and one part belongs to Telkabad village located northeast of Ardestan.

Qanat Dependant Structures

The qanat water civilisation gave rise to a number of traditional structures that supported the ancient hydraulic infrastructure, and many survive and still function today.

Water Reservoirs

The qanats often discharged over great distances into water reservoirs called "*ab anbars*". These reservoirs influenced urban development much as their rural qanats influenced agricultural development. Most of the water reservoirs comprise of a storage tank with hemispherical domed roof, wind tower, entrance and access chamber.

To enable the water to be kept cool and to support the pressure on the walls the main reservoir is buried within the soil. Water reservoirs are often equipped with wind towers (badgir) to cool the water and avoid stagnation. These towers also control the humidity and thus protect the structure. Water reservoirs could provide year round cool water for domestic use. Figure 4.7 shows a water reservoir in Yazd. Figure 4.8

Fig. 4.7 The Hasan Abad Moshir qanat feeds this water reservoir located in a neighbourhood in Yazd, Iran

Fig. 4.8 Water reservoir at Ardestan, Iran showing hemispherical roof and four wind towers (badgirs)

Fig. 4.9 Showing the entrance to the water reservoir at Ardestan, Iran via a sloping tunnel called a "payab"

shows an example of the water reservoir and wind towers at Ardestan. Figure 4.9 shows the entrance to the qanata through a "payab" at street level.

Wind Tower (badgir)

The wind tower (*badgir*) serves as a means of natural ventilation. They were first erected in qanat fed water reservoirs and later adopted to ventilate mosques and private houses. The tower consists of a vertical channel with vertical openings depending on the direction of air currents and the lower portion contains ducts. The difference in height between the base and the top of the tower creates a small pressure difference which allows warm air to travel upwards and fresh cooler air to come down. The airflow is also created by the stack effect due to temperature differences between the inside and outside of the tower. The wind tower is more efficient when there is a larger temperature difference between inside and outside. This natural effect meant that they were very efficient in water reservoirs to create air currents and avoid stagnation (Fig. 4.10).

Payab

A payab is a sloping gallery, connecting the ground surface to the qanat's gallery. There is a pool at the bottom of the payab. This pool is fed by the qanat system. Often a platform was provided for people to rest. In private payabs people used them to store food and meats. Figure 4.11 shows the exterior of a payab in Kashan, Iran.

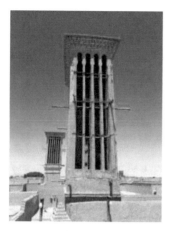

Fig. 4.10 Photo on left shows exterior of a wind tower, photo on right shows ventilation channels in the base of the tower

Fig. 4.11 Photos showing exterior and interior of a payab in a bazaar in Kashan, Iran

4.3   Peru – Puquios

The Nazca region of Peru is one of the most arid regions on the planet. The area is most famous for the intriguing Nazca Lines. These comprise several enormous geometric images carved into the desert sands providing archaeological evidence of the existence of the Nazca people and evidence of the subsequent rapid decline of this once flourishing society. In one of the most arid regions in the world a series of carefully constructed, spiralling holes also form lines across the landscape. Known as puquios, their origin has been a puzzle.

For civilization to exist at all in this region required an ingenious water system that could supply water for agricultural, domestic and ceremonial needs. The puquios system comprised of a series of underground tunnels, surface channels, small reservoirs and most strikingly spiralling holes that lead from the surface to the water below. These systems were designed to retrieve water from underground aquifers and distribute it over many kilometres to where it was needed. Surface channels tended to be shorter than underground ones, and in some cases followed S-shaped curves to slow the water flow. Cobblestones were used to line both the surface and underground channels. Along the course of many of the underground channels are strange spiraling well-like constructions known as "*ojos*" (eyes). On the surface, these ojos can be up to 15 m wide, with their sides corkscrewing inwards as they spiral downwards towards the underground aquifer where the bottom of the well is normally 1 to 2 m wide. Figure 4.12 shows a photo of this spiral structure (ojos).

A study published in *Ancient Nasca World: New Insights from Science and Archeology* in 2006 by Rosa Lasaponara and her team from the Institute of Methodologies for Environmental Analysis in Italy provided new insights on both the ojos and the wider system of puquios.

Lasaponara and her team studied the system using satellite imaging to compare the puquios networks with known Nazca settlements. This gave a better understanding of how the puquios were distributed across the Nazca region. Lasaponara was able to show strong evidence that the Nazca people built the puquios approximately 1,500 years ago.

A study of the ojos system also revealed new information. The ojos had a function beyond that of a well or access point. The spiral shaped holes help to funnel wind down into the underground channels. This helped push the water through the puquios system, the ojos effectively acted like a series of ancient windpumps. Today, there are 36 puquios still functioning in three valleys in the Nazca area. The most famous system is known as the Acueductos de Cantalloc (Cantalloc Aqueducts), located a few miles outside the city of Nazca, where you can see surface channels, underground aqueducts and ojos (Fig. 4.13).

These structures show the native people of the Nazca basin were not only highly organised, but that their society was structured in a hierarchy, says Lasaponara. She says the puquios were a vital tool in controlling water distribution by those in power over the communities that came under their influence. The puquios systems would have required builders who had a deep understanding of the geology of the region.

Fig. 4.12 Photo shows the spiral well like structure (ojos) with their funnel like structure

Fig. 4.13 Photo shows a series of ojos, which functioned as ancient pumps driving water through the underground network of puquios

The construction and maintenance of the systems would have required a collective effort revealing a highly organised and hierarchal society. As Lasaponara concludes *"by knowing how to bring water to one of the driest places on Earth means that you hold the very key of life itself"*. Figure 4.14 shows the systems still in operation today.

Fig. 4.14 Top left photo shows outlet of puquios, bottom centre photo shows view of outlet structure lined both sides with stones, top right photo shows agriculture system still irrigated today with water supplied by the ancient puquios network

4.4 Rome – Impluvium

A compluvium is a space left unroofed over the court of a dwelling in ancient Rome. The impluvium is the sunken part of the atrium designed to carry away the rainwater coming through the compluvium of the roof. It is usually made of marble and placed about 30 cm below the floor of the atrium. The surface of the impluvium was either porous or constructed of tiles with sufficient gaps between them to allow water to pass through into the filter layers below. These filter layers comprised of sand and gravel above an underground cistern chamber.

During periods of intense rainfall any excess water that could not pass through the filter material would overflow the basin and exit the building, and any sediment or debris remaining in the surface basin could be swept away. In hot weather, water could be drawn from the underground cistern chamber and poured into a shallow pool which then evaporated and provided a cooling effect to the entire atrium. As the water evaporated, the surrounding air was cooled and become heavier and flowed into the living spaces and was replaced by air drawn through the compluvium. The combination of the compluvium and impluvium formed an innovative method to harvest, filter, store and cool rainwater. The system made the harvested rainwater readily available for household use as well as providing cooling of the living spaces.

Inspection (without excavation) of impluvia in Paestum, Pompeii and Rome has shown circular stone opening (visible in the photograph, resembling a chimney pot) which allowed easy access by bucket and rope to this private, filtered and naturally cooled water supply. Similar water supplies were found elsewhere in the public spaces of the city with their stone caps showing the wear patterns of much use. Figure 4.15 shows such an example, note the rope wear patterns visible in the photograph.

Fig. 4.15 (**a**) Circular stone opening used to access stored rainwater in underground chamber. (**b**) Impluvia designed to capture and store rainwater

4.5   Chultuns of the Maya Civilisation

The ancient Maya are probably one of the most well-known of the Pre-Columbian civilizations of Central America. The Maya were one of the most advanced civilizations known for their architecture, writing system, calendar, astronomy and mathematics. But among their greatest achievements was also their innovative water management system that made it possible for the civilization to exist on a Yucatan Peninsula with few surface water sources and a six-month dry season.

The Maya water management system relied mainly on harvesting and storing rainwater. The fact that the Maya managed to build a thriving civilization in this environment for 2000 years is testament to their skills as water engineers. However, the success of the systems may have also facilitated a resulting growth in population, which caused the Maya to deforest large parts of the jungle and ultimately made the Maya more vulnerable to droughts leading to the eventual collapse of the civilisation.

Yucatan Peninsula is an area that contains parts of Mexico, Guatemala and Belize. The geology typically comprises of porous limestone covered by a thin layer of soil. Rainwater passes easily through the permeable soil and the porous limestone, and because of this there is nearly a complete lack of surface water on the peninsula. As the rainwater passes through the limestone, some of it is naturally stored in underground caverns. If the limestone on the surface fractures, the water-filled caverns are exposed and form a *"cenote"*. The Maya took advantage of these cenotes as water sources wherever possible.

Where natural caverns were unavailable, the Maya turned to man-made cisterns. These cisterns, knowns as chultuns, were bottle-shaped underground water storage chambers that the Mayans lined with lime plaster to form an impermeable water storage chamber. They then developed a distribution system that allowed users to access the water stored in the underground chambers. The chultuns, stored rainwater in the wet season making it possible for the city to survive periods of prolonged drought.

Thus, in order to survive the Maya had to develop their own water management strategies. Figure 4.16 shows some photos of the chultuna cisterns and includes an

Fig. 4.16 Chulton cisterns used by the Maya Civilisation

illustration of a typical cross section through a chultun showing the narrow opening and a wider, cylindrical body extending to a depth of up to 6m. These chultuns were located adjacent to residences. Some chultuns were found to have water storage capacities ranging from 7 to 50 cubic metres.

4.6   Rainwater Harvesting Solutions in Uganda

Uganda Water Resources
Uganda is a country rich in water resources. The Nile Basin constitutes about 98% of the total area of the country. The whole of Uganda lies within the upper Nile catchments with rivers flowing into Lakes Victoria, Edward, Kyoga, and Albert, and also directly into river Nile. Uganda's rivers and lakes, including wetlands, cover about 20% of the total surface area. Rainfall is the principle contributor of water to the surface water bodies. Groundwater is found in aquifers which are water-bearing formations from which it can be drawn in significant amounts through the use of dug-wells and bore-holes. Uganda's rural population meets its needs from surface water (lakes, rivers and streams), springs and wells, and boreholes. These water sources are of varying quality and availability. Uganda has experienced almost two decades of sustained economic growth which has resulted in large population movements from rural areas to informal urban settlements. Fifty one per cent of Ugandans lack access to safe water and 82 per cent lack access to safe sanitation. Figure 4.17

Fig. 4.17   51% of Ugandans lack access to a safe drinking water supply

shows the daily challenge of accessing water in this environment. With an annual rainfall ranging from 800 to 2500 mm, rainwater harvesting has the potential to provide an alternative water supply for households and communities. Rainwater harvesting has been adopted by the government as a water supply strategy and at household level is subsidised, with householders contributing 60% of the overall costs. Several tank designs are available, with the steel tank popular.

The following example shows two very different approaches to rwh in Uganda. The first example shows the ingenuity of a local farmer who has constructed an extremely low cost system entirely from "bush" resources. The second example shows a proprietary rwh system purchased from a local rwh supplier. The last two examples show communal rwh systems, one in a school and on in a village setting.

DIY Household rwh system This example shows a low cost solution which the householder has constructed from available local resources. The rwh tank comprises an underground tank lined with polyethylene liner, sewed at the seams. The base, sides and top of the tank are lined with timber "bush poles". The gutter has been fabricated from old corrugated iron sheets. The roof catchment is approximately 42 m². With annual rainfall in this location of 800 mm/yr, the estimated yield is 25,200 litres/annum. The tank capacity is an estimated 50,000 litres. The rainwater is not used for drinking, mainly due to the individual householder's fears over poisoning by villagers. The rwh does provide a year round water supply for irrigation and the householders animals. Figure 4.18 shows the homemade rwh conveyance system. Figure 4.19 shows the roof cover of the underground rwh storage tank. Figure 4.20 shows the conveyance system. Figure 4.21 shows an internal view of the underground rwh storage tank.

Fig. 4.18 Home made rwh conveyance system

Fig. 4.19 Roof of underground rwh storage tank

Fig. 4.20 Showing conveyance system to deliver an annual volume of 50,000 litres to the underground rwh tank

Fig. 4.21 The interior of the rwh tank is lined with locally available plastic

Proprietary rwh system   This example shows a locally manufactured rwh system which is fabricated by local technicians and costs approximately 600,000 Ugandan shillings (€240). It comprises an above ground 8000 litres corrugated iron rwh tank. The conveyance systems comprises a series of steel pipes and gutters, also locally manufactured. The rwh tank is placed on a concrete plinth. Harvested rainwater is extracted from a tap at the base of the tank. Figure 4.22 shows this rwh system.

Fig. 4.22 The photos show the conveyance system comprising steel gutter and pipes discharging to a corrugated steel rwh tank

The steel rwh tanks are manufactured locally and are a common site in local houses in this part of Uganda. They provide water for all the households demands (drinking, cooking, washing etc) during the wet season. With careful management they can also supply all the households drinking water demands only during the dry season. In this way they provide a safe reliable household water supply year round. Figure 4.23 shows this particular system.

Fig. 4.23 Shows the above ground rwh 800 litre tank with tap outlet at base

School rwh system   This example shows a rwh system located in a pre-school. The school provides meals to 800 children daily and the rwh system provides water for cooking year round. A series of 10,000 litre water tanks are located at each roof to maximise storage. The conveyance system in this example is a series of 100 mm pvc pipes with pvc fittings. A rwh system also provides water for the school latrines. Figure 4.24 shows this system.

Fig. 4.24 Shows the school rwh system for school kitchen and latrines

Village rwh system This corrugated iron roof shelter was constructed specifically to provide a rainwater catchment area to harvest rainwater for 62 households. Figure 4.25 shows the components of this rwh system. These comprise of a roof catchment discharging directly to a ferrocement rwh tank. The outlet is via an underground tap. The Catchment surface area is 22 m × 10 m (220 m^2) which supplies approximately 150,000 litres/annum, or 410 l/d.

Fig. 4.25 Shows the components of the village rwh system which is used for year round supply of drinking water only

4.7   Sri Lanka Rainwater Technicians

There is not a water shortage in Sri Lanka, the challenge is to manage the available water. Sri Lanka has an average annual rainfall of 2400 mm, with a minimum of 900 mm in the dry zone and up to 5000 mm in the wet zone. Sri Lankan communities have used rainwater for both domestic and agriculture for centuries. Traditional methods included using banana or coconut leaves to collect rainwater, or from rooftops into open barrels and domestic cooking pots. Lanka Rain Water Harvesting Forum (LRWHF) (www.lankarainwater.org) formed in 1997 to promote scientific research into rwh and to develop a sustainable community rwh model. In a collaborative research program with the Development Technology Unit (DTU) at Warwick University (UK) they tested a number of low cost rwh tanks. Outputs from this study are available on the DTU's web portal (www.warwick.ac.uk).

Initially the forum focused on the promotion of two standard technology options: an underground brick tank modelled after a Chinese biogas digester design and a free standing 5 m³ ferrocement tank. The tank was sized on the basis that a family of 5 should have a minimum of 20 l a day for a period of 50 days (dry period). The ferrocement pumpkin shaped tank has since been adopted as the standard design offered.

A number of local masons were initially trained to construct the ferrocement rwh tanks. As shown in Fig. 4.26, the tanks were constructed by fixing a frame tied by 6 mm wire in the shape of a pumpkin. Chicken wire was then tied in a number of layers to the frame. The first layer of cement of 1:3 mixture was then applied to the outside of the mesh to a thickness of about 12 mm. A second layer of plaster was then applied and the structure covered with wet rags and left to dry for one day. A scaffold was then made over the tank and the 6 mm wire and frame removed from the interior of the tank. A 20 mm cement layer of 1:3 mixture was then applied to the interior of the tank. Water proofing additive was applied to this mixture and smoothened. The base and lid of the tank was then plastered. The tank was then left to cure for 14 days after completion.

Fig. 4.26 Construction of 5 m³ ferrocement rainwater harvesting tank

Tanuja Ariyananda of the LRWHF has carried out extensive research on users of the rwh systems. In her studies of rwh systems in Badulla and Matara districts of Sri Lanka only 10% of households use the harvested rainwater for drinking purposes. Perception of the quality of the harvested rainwater and taste seem to dictate what they use it for. Initially many household were leaving the tank open during the rainy season. Subsequent ingress of leaves and insects resulted in bacteriological contamination and many householders reported feeling sick when consuming the rainwater for drinking. Rainwater systems now come with clear instructions for operation and maintenance on the side of the tank as shown in Fig. 4.27.

Fig. 4.27 Shows the components of the rwh system including instructions for use

The tsunamai on the 26th December 2004 made 1.5 million (7.5% of the population) of Sri Lanka homeless. A major reconstruction program with the assistance of the international community was initiated. Many houses were constructed in areas not serviced by municipal water or sanitation. Rainwater harvesting was an important alternative household water supply strategy for these schemes. Figure 4.28 shows an example of the standard rwh system in a tusnamai construction program at Kalutara, Sri Lanka. Note in this design the rwh system is fitted with a first flush device and two gravel / sand filters are fitted to the tank inlet.

Fig. 4.28   rwh system installed in post tsunami houses at Kalutara, Sri Lanka, 2007

4.8   Village Technology Education Centre (V$_{TEC}$), Sierra Leone

The challenges of accessing an adequate quantity and quality of safe drinking water in Sierra Leone cannot be achieved by traditional high cost, externally funded and constructed water and sanitation approaches. Figure 4.29 show the daily struggle for water in Sierra Leone. A strategy which has shown significant improvements across similar countries is based on the concept of self-supply. Self-supply may be defined as the improvement of household or community water and wastewater systems through investment by the user in technology, construction, operation and maintenance. The Technological University Dublin (TU Dublin) in partnership with Engineers Without Borders (EWB) Ireland has developed a methodology for promoting self-supply at household and community level. The vehicle for this is called the "Village Technology Education Centre (V$_{TEC}$ Centre)". These centres use an innovative approach to teaching self supply water and sanitation technologies at household level. As part of an EU funded project a Village Technology Education Centre (VTEC) was established in Genedema to promote sustainable technologies, including rainwater harvesting (www.dit.ie/dtc).

Fig. 4.29 The daily struggle for water in Sierra Leone

Village Technology Education Centre (V$_{TEC}$ Centre)

Without targeted external support, particularly in remote rural areas, households often lack access to information on suitable technical options and there is insufficient capacity in the local private sector to support a self-supply model. The Technological University Dublin (TU Dublin) has developed the "Village Technology Education Centre (V$_{TEC}$ Centre)" as a methodology for promoting self-supply at household level. These centres use an innovative approach to teaching appropriate water and sanitation technologies at community level.

V$_{TEC}$ – Concept

The V$_{TEC}$ is a community water and sanitation resource and training centre which is used to promote a range of water and sanitation technologies. The involvement of the local community is essential and the community will identify a suitable location for the centre. A V$_{TEC}$ centre can be integrated into an existing school, college, agricultural training centre, teacher training centre or other community asset.

V$_{TEC}$ – Portfolio of Technologies

Each V$_{TEC}$ Centre has a technology portfolio of working modular, standardized plug and play demonstration water & wastewater systems. These serve as tactile teaching units where participants learn how to design, construct, assemble, operate and maintain the technologies. The program is innovative in its educational approach but solutions are based on proven scientific and engineering technologies. A typical V$_{TEC}$ portfolio will include:

Rainwater Harvesting
Solar Water Disinfection Systems
Natural Wastewater Treatment & Reuse Systems
Agricultural Water Systems
Low Cost Water Quality Testing

The whole approach can be summarized as standardized, modular, plug and play. The philosophy is that all technological parts are standardized, i.e. of a specified type and size and these standardized parts are based on local availability. Therefore, all parts of the systems will be available locally. The same parts will be used in all locations. This will both create a market for these parts but will also ensure that these parts are available and the technologies will continue to function. The technological systems themselves be they rainwater harvesting systems or natural wastewater treatment systems are made up of different modules. This facilitates maintenance. If a module fails due to age or overuse, a replacement module can be installed without compromising the whole system. This modular system will also facilitate scaling individual household systems to village systems, if required. The concept of plug and play allows for the set-up of the technology and its use without user involvement in commissioning or designing the technological product. The individual components are simply put together, "plug" and are instantly fit for use "play". Figure 4.30 shows an example of the tactile demonstration units.

Fig. 4.30 Shows tactile working demonstration units for rainwater harvesting, solar water disinfection, latrine slabs and low cost canzee pumps for rwh systems

V$_{TEC}$ – Training for Trainers

The community initially identifies the trainees. These can be local teachers, skilled trades people, or other members of the community. These will be trained in the centre initially by TU Dublin personnel. They will then have to construct the technologies in a local household. Once the installations have been inspected and certified they are then certified as approved V$_{TEC}$ trainers. They then become "trainers of trainers". Figure 4.31 shows trainers on a course.

Fig. 4.31 Shows students during the practical skills training courses

V$_{TEC}$ – Higher Education Involvement

Each centre has a higher education involvement through TU Dublin. TU Dublin have developed a number of modules focused on teaching the science and engineering of water, wastewater and water quality testing. Delivery is through innovative educational pedagogical approaches which have been developed by TU Dublin to bypass the traditional literacy challenges. These include such elements as technical comics, audio visual lectures designed to be delivered using low tech raspberry pi modules suitable for village settings and hands on demonstrations.

V_{TEC} – Community Evaluation and Monitoring System

An innovative feature of the V_{TEC} concept is the inclusion of the element of experimental testing and monitoring of the technologies installed in the community. Using scientific and engineering methods and instruments, the TU Dublin team has produced a series of resources to enable all who have been trained in the centre to gain a practical knowledge of the measurement and monitoring of scientific and engineering parameters relevant to water and sanitation. The skills and the equipment will be made available through the centre in order to facilitate accurate monitoring of the water quality produced by each technology. This will promote a greater understanding of the removal mechanisms involved in the appliances installed but it will also serve to reinforce the applicability of the technologies and to show the success, or indeed failure, of the various installations. The importance of this section of the V_{TEC} cannot be stressed enough. It allows the community to monitor their own environment while reinforcing the concepts of basic sanitation. This is a very important and novel part, of the V_{TEC} concept as promoted by TU Dublin.

V_{TEC} – Training for Trainers

Students quickly learnt the skills and started to reproduce and replicate the designs in the resource centre workshops. Local students were trained as V_{TEC} trainers. These trainers ran courses for local villagers and small scale entrepreneurs. They visited the villages promoting the technologies and assisting the villagers with supervision of the installation of the various self-supply applications. Villagers were encouraged to visit the technology centre in the school and take apart and play with the technological exhibits on display. This hands on experience created a greater understanding of the technologies for the villagers and helped them comprehend construction problems and system characteristics. They were also trained in scientific monitoring of water quality. Figure 4.32 shows students learning water testing methods. Figure 4.33 shows students installing working examples of the technologies.

Fig. 4.32 V_{TEC} students learning to monitor water quality in the household systems

Fig. 4.33 V_{TEC} Students building full scale rwh systems at the training centre

V_{TEC} – Small Medium Enterprises (SME) Involvement
An important element with the self-supply model is building up the capacity of local entrepreneurs.

Anyone who completes a course in the centre and who has demonstrated a fully functioning system is certified as an approved V_{TEC} Technologist. In order to maintain this certification they are required to undergo annual accreditation which will be obtained through attendance at and graduation from, continuous professional development (CPD) courses at the centre. This is developed in conjunction with local educational institutes. This initiative will have a focus on the creation of small and medium enterprises (SMEs) as parts and spare parts will be required to maintain and install the technologies. The market created will allow SMEs to flourish. Figure 4.34 shows components manufactured ready for installation in villages.

Water storage tanks fabricated by the students in the VTKC are tested for quality and water tightness before installation.

Fig. 4.34 Each V_{TEC} centre becomes a production centre for components within the village rwh supply chain

Household Water Supply System (Fig. 4.35)

Fig. 4.35 Example of a household rwh system installed by a V_{TEC} graduate comprising standardized, modular, plug and play rwh components fabricated using locally available parts within the V_{TEC} centre

4.9   Angel House

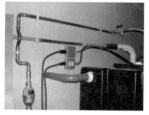

Fig. 4.36   RWH system at Angel Town, London Borough of Lambeth

The Angel Town estate in the London Borough of Lambeth was built in the early
1970's and consists of four storey deck access blocks linked with high-level bridges
and garages at ground level. In the early 1990's the bridges connecting the blocks
were demolished and a pilot scheme was set up to convert the first blocks. Security,
energy conservation and measures to increase the levels of comfort were prioritised
in the refurbishment of Holles House. The aim of the project was to meet sustainable objectives in the context of regeneration. A comprehensive package of energy
improvements to the fabric of the building and the heating and hot water systems
was developed. Rainwater harvesting was incorporated into the scheme and harvested rainwater from the roof was distributed to each flat via individual rwh control
systems as shown. This work took place at the same time as improvements to the
local environment to improve safety and security and amenities. Residents were
involved in decisions for the refurbishment programme, and the Estate Group had
close links with the architect throughout the project. Residents were on the panel
that chose the architects and contractors.

4.10   Ultra Pure Water: Water from Heavan

Drinking water from own roof

Water from Heaven – Hemel(s) water® – Preciouspitation is very pure water made out of rainwater. It is produced sustainably using an Ultra Filtration membrane.

Introduction
The world is faced with a scarcity of (fresh) water while at the same time extreme rainfall occurs. This water crisis has led the World Economic Forum, since 2015, to make water the primary threat to the economy. Currently 600 million people do not have access to drinking water and a majority of the world's population lack safe drinking water (UN).

Solution
This technical solution developed in the Netherlands is simple and new. Rainwater is harvested from the roof and is treated by collecting, purifying and conditioning to prevent evaporation and growth of bacteria or algae, to achieve potable water standards. The whole system works by gravity so no power or chemicals are required.

Technology (Patent Pending)
In the upper tank storm water is collected from the roof. After passing a denutritor (removing NH_4^+) and membrane filtration (removing bacteria and viruses) by gravity, the water is purified and stored in the bottom tank.

Awards Winning Concept
Water from Heaven, winner of the Challenge City of the Future 2016 organized by 3 Dutch ministries; Finalist of the Herman Wijffels Innovation award 2016 and of the Dutch Water innovation prize 2017.

Value Drivers
Safety: Rain harvested and stored at warm temperatures ($\gg 10\ °C$) can lead to bacterial growth. Water from Heaven is bacteria free (Legionella, Zika, Malaria, Ebola, Guardia, Enterococci, etc.).

   Health: There are strong concerns on increasing concentrations of hormones and medicines (EDCs) in drinking water. Rainwater is free of these because it is distilled water.

   Sustainability: *Water from Heaven* is produced without CO_2 emission. It uses no power nor chemicals and leaves no waste streams.

   Fresh water scarcity: Rainwater is normally spilled to river/sea/ground or evaporated. For this reason, we now have to drink treated surface- or groundwater; in fact, unnecessary contaminated rainwater.

   Climate resilient cities: Rainwater harvesting contributes to rainproof streets and houses.

Social advantages: Less fresh water to the sewage system increases the concentration of waste in urban waste water, which increases the treatment efficiency.

In areas of non-drinkable water sources this opens possibilities to prevent long distance water collection or high energy intensive treatment.

Independence: Each house with its own drinking water facility makes civilians independent of non-reliable or non-existent municipal supply (4 billion people consume drinking water that does not comply with WHO guidelines)

Water Quality Results

The first pilot studies in 2015 showed good water quality. The results show that concentrations of all chemicals are very low compared to drinking water, except ammonium. After Ultra Filtration *Water from Heaven* is completely bacteria and virus free. In the Brigaid project it is shown that a Denutritor removes NH_4^+ adequately. In the industrial pilots it is found that conductivity maybe as low as 20 µS/cm and no pesticides were present.

Drinking Water Standards

The current water quality in the tested pilot studies meet several drinking water standards (NL-EU-WHO). Sometimes ammonia or colony counts were in excess of the standards. All other contaminants were far lower than the standards. Therefore, the pilot partners from the Netherlands have proposed a new standard for harvested rainwater that can achieve a higher water quality than the current Drinking Water Standards. The proposed new harvested rainwater standard requirements are equal to drinking water standards but much stricter (NO_3^-, Ca^{++}, Na^+, Cl^-, conductivity) and there are even new requirements for medicines, EDC's, PAK's plastics an nanoparticles etc. This standard is under development in the Netherlands as is the certification process.

Market Applications

This high quality water can be applied EU wide and is now promoted to be tested in each country separately. It can be used as drinking water, industrial water (food and non-food) or agro-water (feedwater or dripping water).

Costs

It is forecasted that the total cost of ownership of a heavenly water installation equals to current drinking water costs. In the future, this could be lowered by introduction of mass production and a do-it-yourself approach.

R&D

This technology is currently researched in an EU project (Brigaid) by Catholic University of Leuven and in a Dutch Environmental Protection Agency project to study possibilities to use the harvested rainwater as process water in industry to improve sustainability in the long term at several Dutch industries.

More information: Albert.wic@ziggo.nl

4.11   Rainwater Harvesting Systems in Ireland

Domestic rainwater harvesting system installed in Carlow, Ireland.   This is an example of an indirect gravity system. Down pipes and guttering were connected via an inline Rainman Type1 filter to the inlet of the storage tank in the garden. Figure 4.37 shows the rwh system. A calming inlet in the storage tank was installed

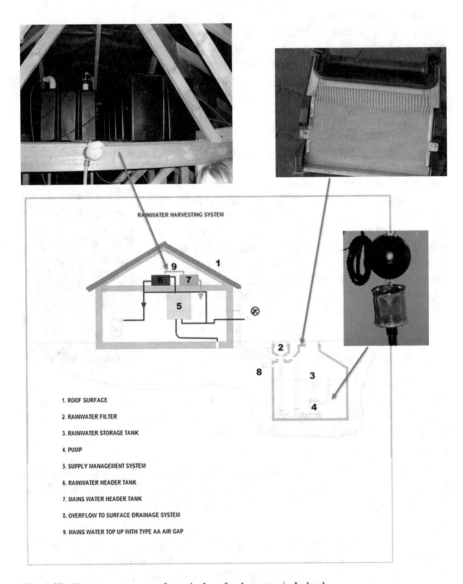

Fig. 4.37   Shows components of a typical roof rwh system in Ireland

which calmed the water as it entered the tank preventing any settled material at the tank bottom being disturbed. To ensure that the rainwater quality was not compromised by debris and dirt from paths around the house, points where the down pipes connect with the ground pipes were sealed using silicone sealant. This seal prevented dirt inflow to the RWH system and was easily removable if required. In addition to the normal mains header tank in the attic, a second header tank was installed. The mains header tank supplied the hot and cold water systems excluding the toilets. The extra header tank provided storage for the rainwater supplying the toilets. A submersible Multigo pump placed in the storage tank in the garden pumped the collected rainwater to the rainwater header tank in the attic. The pump's floating filter inlet lay just below the water level, preventing any floating debris entering the pump. The pump had a safety mechanism which prevented the pump switching on if the water level in the tank was below a certain level. This protected the pump and prevented any settled material being disturbed, thus clogging the pump inlet or entering the rainwater header tank.

Agricultural Rainwater Harvesting System   This is an example of an indirect gravity system. Figure 4.38 shows the system. The guttering and down pipes on the farm buildings were replaced. Three 100 mm down pipes were used to convey the rainwater under gravity from the buildings' roofs to the rwh system. Harvested rainwater is conveyed via underground pipe work to a 9 m³ collection tank A submersible Multigo pump was installed in the collection tank to allow pumping of rainwater up to the storage tank. From the collection tank the rainwater was pumped up to two interconnected 22 m³ concrete storage tanks giving a total storage capacity of 44 m³. The cattle troughs on the farm were fed from the storage tank by gravity feed. The storage tank is approximately 10 m above the collection tank and the harvested rainwater fed cattle troughs.

School rwh system   This is an example of a gravity system. Figure 4.39 shows the system layout and components. The rwh system was divided into two separate systems installed on either side of the building. The roof catchment area draining to each system was 62 m², providing a total catchment area of 124 m². The roof covering a Sarnafil® membrane. The roof is pitched at approximately 10° resulting in rainwater draining to the roof edge which comprised a 150 mm deep perimeter. Within the roof perimeter gutters channel the rainwater directly to the storage tanks located on each side of the roof space. Each tank has a storage capacity of 682 litres and is fitted with a lid. Rainwater is channelled from the external channel to the internal storage tank. Rainwater is directed straight through to the storage tank via a 50 mm pipe. Attached to this pipe is a cloth geotextile filter with 4 mm aperture to capture any fines from the roof surface. A 100 mm diameter overflow pipe drains directly to the external roof channel. Mains water is piped to the rainwater storage tank and controlled by a ball cock valve. The level of the ball cock is set below the level of the intake from the rainwater. This top up system ensures reliability of supply. A secondary overflow system from the storage tank comprises two 50 mm pipes. There are no pumps in this system. Rainwater from the roof drains by gravity to the storage tanks. Supply from the storage tanks to the building toilets is a gravity system.

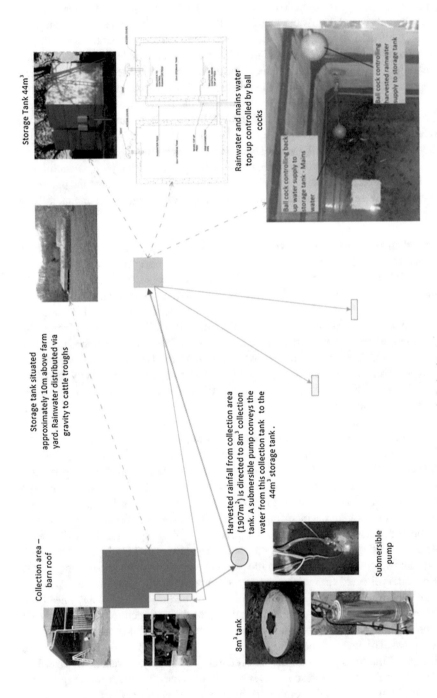

Fig. 4.38 Shows the components of an agricultural rwh system in Ireland

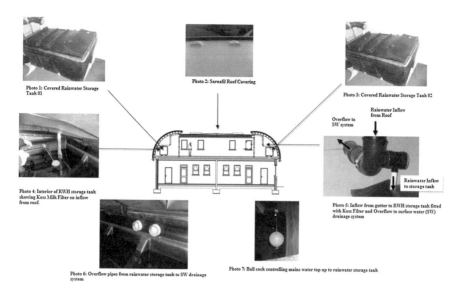

Fig. 4.39   Components of a school rwh system, Ireland

4.12   Singapore: The Role of Rainwater Harvesting in a Resilient Water Supply System

Singapore is an example of the holistic approach to water, on the level of a nation state. All hard surfaces are considered water catchments, and all water that falls is considered a useful resource. These unprotected water catchments supply raw water which is treated to potable water standards. Wastewater is recycled and treated to drinking water standard and resold to users as a high quality brand "NEWater".

Singapore is an example where innovation in thinking has resulted in it moving from being a net importer of water to becoming self-sufficient in a period of 30 years. At the time of independence, Singapore was dependent upon importing water from its neighbour, Malaysia, to supplement water supply. This led to a policy decision in the Prime Minister's office "that water should govern every Government decision".

Singaporeans refer to the water loop that is the basis for a sustainable water supply as the "Four Taps". Figure 4.40 illustrates this strategy.

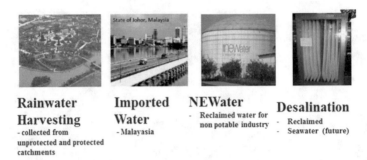

Rainwater Harvesting	Imported Water	NEWater	Desalination
- collected from unprotected and protected catchments	- Malayasia	- Reclaimed water for non potable industry	- Reclaimed - Seawater (future)

Fig. 4.40 Singapore Four Taps water management strategy

The First National Tap refers to rainwater harvesting that supplies raw water that is treated to produce potable water. The harvested rainwater is collected from a network of waterways and reservoirs throughout the peninsula. These were initially fed by protected catchments, areas of land where industrial and housing developments were strictly controlled to protect the quality of the rainwater which was harvested. However, as land demands increased for housing and industry, less land could be spared for additional protected water catchments. The innovation in thinking was to consider harvesting rainwater from unprotected catchments. Unprotected catchments are areas of water catchment where all types of land use are allowed upstream of the storage area, regardless of potential effects on water quality. These include parks, pavements, roads, drains etc. Every drop of rain is harvested and all surfaces are regarded as water catchments. Water is designed into a system, as against being designed out via storm water drainage. Figure 4.41 illustrates this strategy. Figure 4.42 shows a typical "unprotected" water catchment in an urban setting.

"Two thirds of Singapore is already a water catchment"

Rainwater harvested on-site

Rainwater channelled to canals

Discharged to Reservoirs

Fig. 4.41 Rainwater is harvested from all impermeable surfaces in Singapore

The Second National Tap refers to the imported water from Malaysia, which was the main source of water prior to the 1960s and is still purchased, but no longer has the same strategic value.

The Third National Tap is NEWater In 1998 two Singaporean engineers were sent on a study trip to the USA, specifically Southern California and Florida. This

Fig. 4.42 Shows a typical "unprotected catchment" in Singapore where rainwater is harvested

trip was a turning point in Singapore's efforts to recycle its wastewater. Wastewater is now treated to a potable standard and the end product is branded as "NEWater" and sold to Information Technology companies and also used in the national water supply. Figure 4.43 shows the "NEWater" facility.

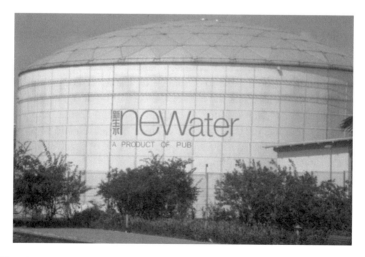

Fig. 4.43 "NEWater" all wastewater is now considered a resource and is treated to a potable standard in Singapore

The **Fourth National Tap is Desalination**, which is being developed with the focus on reducing the power inputs required.

These initiatives were not without their difficulties. The engineering profession opposed the introduction of unprotected catchments on the grounds that water quality would be compromised and that drinking water standards would prove impossible to achieve. To overcome consumer reluctance to accept the concept of drinking water produced from sewage—literally their own excreted waste—a public campaign was conducted to convince consumers of the potability, lack of taste and odour of this treated water. This was rebranded "NEWater" to offset the "yuck factor". The campaign culminated with a high-visibility event at the 2002 National Day Parade when the then Prime Minister Tong lead 60,000 people in a toast to Singapore with "NEWater" as the beverage.

More recently the Singapore government launched a campaign that they call ABC, referring to an Active Beautiful and Clean waters programme. This is an initiative aimed at improving the quality of water and life by harnessing the full potential of waterbodies. By integrating the drains, canals and reservoirs with the surrounding environment in a holistic way, the ABC Waters Programme aims to create beautiful and clean streams, rivers, and lakes with postcard-pretty community spaces for all to enjoy. The idea is to promote local people's participation in, and use of, the waterways of Singapore in the hope that this will instill appreciation and water values into the community. Singapore is promoting the Worth of Water. As they say themselves, "we used to keep the community away from our water, now we want them to use it, play in it, respect it. In other words, to take ownership of water and water resources".

4.13   Rain Cities of South Korea

Rainwater Management in Seoul City
The rain cities of South Korea are an example of rainwater harvesting applied at city level.

Faced with the consequences of climate change, increased flooding and water supply problems, the South Korean Ministry of Land, Transport and Maritime Affairs (MLTM) announced that the rain cities were to be designed such that rainwater is collected rather than running to surface/sewerage drains. This policy of a multi-purpose system has the dual purpose of flood mitigation and water conservation. A special feature of the rwh systems is the provision of a network for monitoring the water levels in all rwh tanks by the central disaster prevention agency. Depending on the predicted rainfall, the agency may issue an order to building owners to empty their rwh tanks. This facilitates a city wide buffer capacity to cater for initial flood storage in the event of intense rainfall events. This initiative led to the adoption of the Rain City policy by many cities throughout South Korea.

Star City, Seoul Star City was a major development project in Gwangjin-gu, a district in eastern Seoul. It consisted of more than 1300 apartment units. The design team for the complex, uniquely, comprised of an alliance of academics, the land owner and developer, the local government, the designer of the development project, and the general contractor. The Star City rainwater harvesting project was designed to capture the first 100 mm of rainfall on the complex and to use the harvested water for toilets and gardening. Three 1000 m³ rwh tanks were installed in the basement. The first tank collects rainwater from unpaved surfaces. This is normally kept empty and used only in intense rainfall events. The second tank collects rwh from the roof, which is used for toilet flushing and landscape purposes. The third rwh tank is kept full and only used for supply during emergencies or firefighting. This project was awarded the International Water Association (IWA) 2010 Project Innovation Award. Results from the Star City project show a local rainwater utilisation rate of 67%. This is a measure of the amount of rainwater used compared to the total annual rainfall in the area. Therefore 67% of the annual rainfall was captured and utilised.

Rainwater Piggy Banks and Micro-credit

Han Mooyoung from the Department of Civil and Environmental Engineering, Seoul University in a paper in the Environmental Engineering Research Journal June 2018, has detailed some interesting examples of pro-active rainwater harvesting management in Korea. A rwh program to encourage the installation of individual household rwh systems was implemented in Seoul City and Suwon City. The system consists of a downpipe rwh filter, a 400–1000 litre rwh tank (referred to as a "piggy bank"), a water meter and an optional infiltration box. The rwh installation costs are covered from a 50% subsidy from Seoul's Metropolitian Government and a 25% subsidy from sponsorship by private firms. So far 59 cities in Korea have passed regulations to subsidize similar rwh facilities. Figure 4.44 shows an example of one of these systems.

Fig. 4.44 rwh "piggy banks" in Seoul

Chapter 5
Rainwater Harvesting Systems

NO ONE VALUES RAINWATER!

5.1 Introduction

Rainwater harvesting (rwh) may be defined as the collection and storage of rainwater falling on impermeable surface(s) for later use. Common to any rwh system are the following components;

CATCHMENT SURFACE	CONVEYANCE SYSTEM	FILTRATION SYSTEM	STORAGE

© Springer Nature Switzerland AG 2021
L. McCarton et al., *The Worth of Water*,
https://doi.org/10.1007/978-3-030-50605-6_5

Catchment Surface

The catchment surface is an impermeable surface used to collect rain falling on it. Most rainwater harvesting systems (rwh) use a roof as the catchment area. Roof harvested rainwater is often utilised for non-potable uses within the house. With appropriate treatment it can also be utilised for potable (drinking) water. Where rainwater is harvested from the ground, it is often utilised for agricultural irrigation and aquifer recharge.

Conveyance System

The rainwater collected on the catchment surface may be stored for later use. A conveyance system allows rainwater collected on catchment surface to be diverted to a storage facility. Conveyance systems can vary from simple earth channels to pvc guttering.

Filtration Systems

The purpose of a filtration system is to remove suspended particles within the harvested rainwater and to prevent them entering the storage system.

Storage System

Storage systems vary widely in type and complexity. In most cases, a tank is used to store harvested rainwater. Tanks are constructed from a variety of materials including concrete, plastic, earthenware etc.

The following sections discuss the main components within a roof rwh system. Later sections discuss some ground rwh systems.

5.2   Roof Rainwater Harvesting Systems

Roof rainwater harvesting refers to the collection and storage of rainwater falling on roof surfaces only. The area on which the rain falls is called the "**catchment area**". The actual area from which the rainwater is harvested and conveyed to the storage system is called the "collection area". Figure 5.1 illustrates this concept.

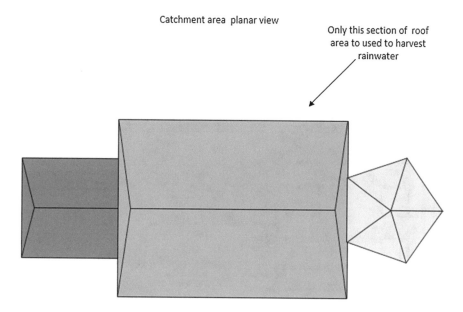

Catchment area planar view

Only this section of roof
area to used to harvest
rainwater

Roof of the building is the catchment area, as rain falls on the total area of
the roof. However in this example only the section coloured green is to be
used to collect rainwater and convey it to the storage system. This green
section is the RWH collection area.

Fig. 5.1   Roof rwh system catchment and collection areas

5.3   Water Flows Within the RWH Process

Rainfall on the roof (collection area) flows to the gutters and downpipes (conveyance system). The rainwater passes through either a first flush device and or a filtration system. The treated harvested rainwater flows to the rainwater storage
tank. The harvested rainwater is delivered to the point of use. A back up water
supply to the rwh system is typically provided to meet rwh demand during periods
of insufficient rainfall. This back up supply system is designed to prevent cross
contamination of potable water supply. Figure 5.2 shows the water flows in a typical roof rwh system.

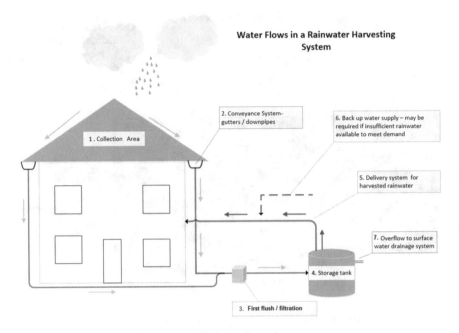

Fig. 5.2   Schematic of water flows in a typical roof rwh system

5.4   RWH Components

Rainwater harvesting for the purpose of this discussion shall be confined to rain collected from roof surfaces only. Rainwater collected from ground surfaces is discussed later in this chapter. Rain falls onto roofs and then runs off. The run-off is variable depending on local climate and rainfall patterns. Runoff from the roof surface can be channelled into a storage tank. Water can be drawn from that store whenever it is needed, hours, days or even months after the last rainfall. The proximity of the water storage to the household can eliminate the need for water to be carried or piped from a distant water source.

A typical roof rainwater harvesting system consists of a number of basic components;

- collection area;
- conveyance system;
- first flush and filter device(s);
- storage tank; and
- delivery system to transport the rainwater to the point of use.

Collection Area

If the rwh is intended to supply or augment potable supply, careful consideration must be given to the material used for the roof. Certain roofing materials such as asbestos cannot be used where harvested rain is to be used for drinking water supply.

Conveyance System

The conveyance system consists of gutters, and downpipes which discharge to a rwh storage tank. Roof outlets, guttering and pipe work should allow for routine maintenance and cleaning. Collection pipe work should allow the rainwater to flow by gravity or siphonic action to the storage tank(s). To prevent stagnation, the pipe work should be free draining and prevent contaminated water entering the system from outside sources. Gutters must be kept clean and free flowing. Figure 5.3 shows examples of gutters with a lack of maintenance.

Ground level gullies should be sealed to prevent debris/pollutants entering the system. Figure 5.4 shows a downpipe sealed at ground level.

Fig. 5.3 Gutters must be kept clear and free flowing

Fig. 5.4 Downpipe sealed at ground level

Fig. 5.5 Roof rwh system at Eden, UK

Fig. 5.6 Mesh is often fitted to prevent debris building up in the gutter

Guttering can be either standard construction material or specially designed to either maximise the amount of harvested rainfall or to enhance the building facade. Figure 5.5 shows the roof rwh system at Eden, UK. Figure 5.6 shows a mesh used to prevent debris build up in the gutter.

Using gutter guards or mesh guards to prevent debris and leaves entering the downpipe reduces the pollutant load on the water in the storage tank.

First Flush and Filters
Protecting the quality of the harvested rainwater requires that roof debris and dirt be excluded from entering the storage system. This can be achieved by installing a first flush device and or a filter system.

First Flush Diversion
Collecting the first few millimetres of rainfall and discarding it protects the rainwater quality entering the storage tank. This is known as the first flush. Any accumulated debris or pollution sources present on the roof surface are removed before they can

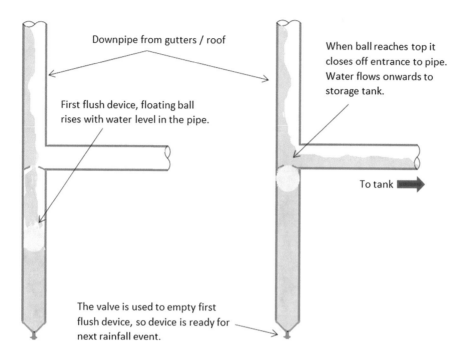

Fig. 5.7 Example of a typical first flush device

enter the storage tank. The World Health Organisation (WHO) recommends the first 20–25 litres for the average sized roof be diverted and discarded. One example of a first flush device is given in Figure 5.7. As rainwater enters the chamber, it fills, and the float rises with the water level until the float closes off the entrance to the chamber. The rainwater then flows to the storage tank. The chamber holds the first rainfall and any debris washed from the roof. This chamber has a small opening that slowly releases the water to waste, resetting the first flush device for the next rainfall event.

Filtration System
A filtration system typically removes waterborne particles upstream of the storage tank. A filtration system should include a filter that has the following characteristics;

- water and weather resistant
- removable and readily accessible for maintenance purposes
- has an efficiency of at least 90%
- passes a maximum particle size of <1.25 mm.

The type of filtration required depends on the ultimate use of the harvested rainwater.

A floating filter may also be used in conjunction with a submersible pump in the storage tank. This facilitates pumping of water from below the surface of the water in the tank. This reduces the likelihood of any floating debris entering the system. Figure 5.8 shows the filtration components of a typical roof rwh system.

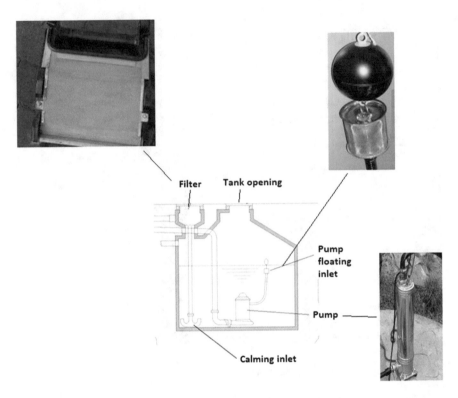

Fig. 5.8 Filtration system components

Storage

The storage system consists of a tank or tanks which store the water for immediate and future use. The inlet to the tank from the conveyance system should be via a calmed inlet. The velocity of the water is slowed down so that it enters the tank without disturbing and re-suspending any fine settled material or biofilm. The calmed inlet facilitates the introduction of oxygen to the bottom layers in the tank by introducing aerated water to the bottom of the tank. Pipework connections should allow the through flow of water to avoid problems of water stagnation, ventilation points should be screened to prevent ingress of vermin. Tanks should have lids to prevent contamination. Lids must be securely fitted to prevent drowning hazards and to omit light thus preventing algal growth. For tanks positioned above ground risk of leakage must be considered, hence additional drainage and bunding may be necessary. Loading of the structure is important when deciding on a location for storage tanks. Figure 5.9 shows these elements in a rwh storage tank.

Back Up Water Supply

In the event of periods of insufficient rainfall, a back-water supply is required to meet demand. It is most important that public water (mains water) is protected from any possibility of contamination from rainwater. A type AA air gap device is used to ensure that there can be no suction or backflow of harvested rainwater into the mains water system. Figure 5.10 shows an illustration of this type of device.

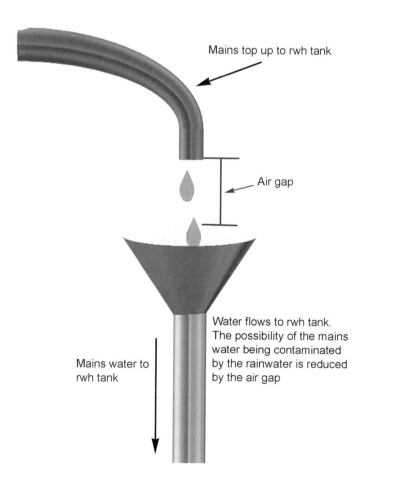

Fig. 5.9 Calming inlet to rwh storage tank

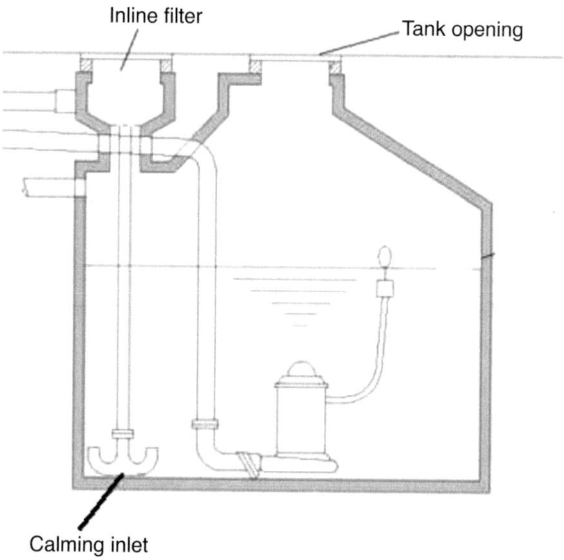

Inline filter

Tank opening

Calming inlet

Mains top up to rwh tank

Air gap

Water flows to rwh tank. The possibility of the mains water being contaminated by the rainwater is reduced by the air gap

Mains water to rwh tank

Fig. 5.10 Example of a Type AA air gap backflow prevention device

5.5   Delivery System

There are three basic types of rwh delivery systems used to supply harvested rainwater water to the point of use. They are described and illustrated as follows;

 (i) **Direct rwh system** – harvested rainwater pumped directly to the points of use;
 (ii) **Gravity rwh system** – harvested rainwater fed by gravity to the points of use;
(iii) **Indirect rwh system** – harvested rainwater pumped to an elevated cistern and fed by gravity to the points of use.

Direct rwh System

In a direct system, rainwater is collected from the roof collection area and conveyed to a storage tank. The harvested rainwater is then pumped directly from the storage tank to the point of use. Figure 5.11 shows a direct rwh system.

Mains top up is provided to the storage tank to ensure constant supply of water for use in the event of long periods of dry weather. No header tank is used in this system. When the storage tank is full any incoming harvested rainwater will overflow to the surface water drainage system. The main disadvantage of directly pumped systems are that in the case of pump failure or power failure no water can be supplied to points of use.

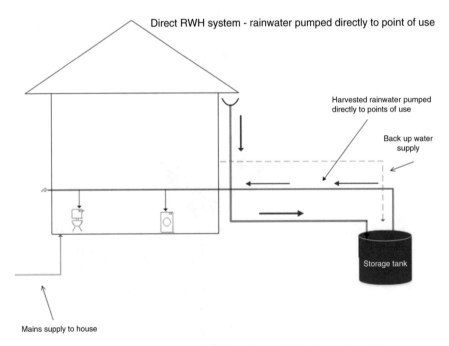

Fig. 5.11 Direct rwh system – rainwater collected and pumped directly to points of use

Gravity rwh System

In a gravity fed rwh system, the harvested rainwater is collected, filtered and conveyed directly to the storage header tank in the roof space or attic. The harvested rainwater is delivered to the points of use via gravity. Mains top-up is supplied to the storage header tank to ensure water availability in the event of a dry weather. To prevent non-potable water entering the potable or public water supply, the back-up water supply should be fitted with a Type AA air gap protection unit that is capable of providing protection against contamination. When the rwh storage tank is full, any additional rainwater supplied to the tank will overflow to the surface water drainage system. An advantage of this system is that no pump is required unlike the other systems, so electricity supply is not required, and risk of pump failure is eliminated. The main disadvantage of a gravity system is the limited rwh storage volume. Figure 5.12 shows a typical gravity rwh system.

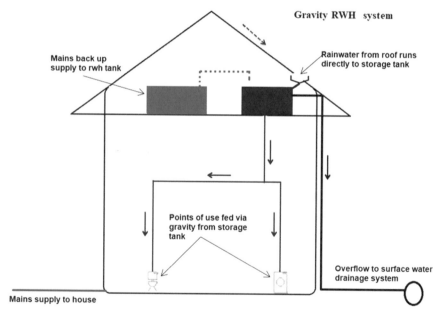

Fig. 5.12 Gravity rwh system – rainwater is collected in a storage tank and distributed by gravity to the points of use

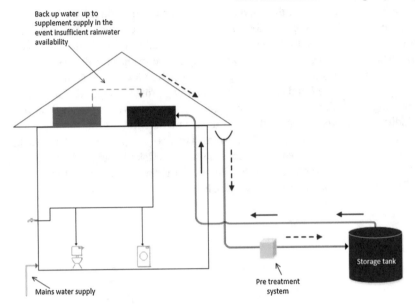

Water flows schematic for Indirect gravity RWH system

Fig. 5.13 Indirect rwh system harvested rainwater collected in storage tank, pumped to an elevated tank and fed by gravity to the points of use

Indirect rwh System

In an indirect rwh system, the harvested rainwater is stored in an underground storage tank, and then pumped to an elevated tank. The points of use are supplied by gravity. Mains supply acts as a backup in the event that there is insufficient harvested rainwater available. In this rwh system, in the event of pump failure water can still be supplied to the points of use via the mains top-up facility. This system has the advantage that if there is a pump or power failure toilet demand is supplied and toilets can be flushed normally. Figure 5.13 shows a typical indirect rwh system.

5.6   Ground RWH Systems

One of the main alternatives to roof rainwater harvesting is to capture the rainwater and utilise it for artificial recharge to the groundwater and / or agricultural irrigation. There are a number of types of ground based rwh systems.

Lined Underground Reservoir

It is a hole dug in the ground, used to collect and store surface runoff from uncultivated ground. It is typically used to harvest rainwater for livestock and / or agricultural irrigation purposes.

Contour Ridges

Contour ridges, sometimes called contour furrows or micro watersheds, are used mainly for crop production. Ridges follow the contours at a spacing of usually 1–2 m. Runoff is collected from the uncultivated strip between ridges and stored in a furrow just above the ridges. Crops are planted on both sides of the furrow. These harvest the rainwater to conserve soil moisture for crop production and also to reduce erosion.

Contour Stone Bunding

In this technique a stone bund is laid along a contour. The contour stone bunds do not concentrate runoff but keep it spread. They also reduce the rate of runoff allowing infiltration. This technique is suited to semi-arid regions where stones are available. The stone bunds do not readily wash away and, therefore, the technique is not vulnerable to unusual and variable intensity rainfall events.

Terracing Contour Bunds

Terracing contour bunds are ridges and ditches made of soil, dug across the slope along the contour. They are used to prevent run-off and to conserve soil and water. Crops are planted on the land between the bunds. The bunds can be stabilised with planting. The distance between the bunds depends on the slope but may be up to 20 m apart.

Permeable Rock Dams

Permeable rock dams consist of long, low rock walls with level crests along the full length across valley floors. This causes runoff to spread laterally from the stream course. This is a floodwater harvesting technique. They are used to retain floodwater for improved crop growth. They can also result in improved land management as a result of silting up of gullies with settled sediment from the floodwater.

Recharge Pits/Trenches

Recharge pits and trenches are constructed for recharging the shallow aquifers and / or avoiding runoff damages. Pits are generally 1–2 m wide and 2–3 m deep. Trenches are generally 0.5–1 m wide and 1–1.5 m deep and 10–20 m long depending upon availability of water.

Both are filled with boulders, gravels & coarse sand to filter and increase water infiltration and to minimize evaporation loss.

Check Dam

A check dam is a small, temporary or permanent dam constructed across a drainage ditch, swale, or channel to lower the speed of concentrated flows for a certain design range of storm events. A check dam can be built from logs of wood, stone, pea gravel-filled sandbags or bricks and cement.

Chapter 6
Health Effects of Utilising RWH

6.1 Introduction

In Europe, the United States of America, Asia, Africa and Canada public water supply is treated to potable standard by various forms of treatment. Water and Government authorities in these regions see untreated rainwater as a probable source of disease and consider rainwater as a source of contamination to potable supplies. As a result, there are stringent legal standards and guidelines with regard to the design of rainwater harvesting systems aimed at avoiding any probability of harvested rainwater entering the public water supply system or being used as potable water. Harvested rainfall (rwh) is typically only used for non-potable uses. Non-potable water is defined as water which does not meet drinking water quality in accordance with Council Directive 98/83/EC.

However, harvested rainwater is used as the main source of household water, including drinking water, in some developed countries. The most notable of these countries is Australia. It is estimated that 10% of Australian households use rwh as their main source of drinking water. In South Australia it is estimated that 45% of households had a rwh system with 22% using them as the main source of drinking water. Hot water systems represent approximately 20% of household per capita consumption and represent a logical extension for the use of rwh to reduce mains water demand.

This chapter sets out to review the international evidence to determine whether using rwh as a main household water sources presents an increased public health risk over mains water supply only systems and to investigate if hot water systems supplied with rwh present an increased risk over hot water systems supplied with mains water only.

6.2   Health Concerns

The most common source of pathogens in drinking water supplies is recent contamination by human and/or animal excreta (Prescott et al. 1993). These include bacteria, viruses and protozoa which are sources of gastroenteritis, diarrhoea, dysentery, hepatitis, cholera or typhoid fever (NHMRC 1996). While the majority of waterborne diseases are caused by pathogens that originate in the gastrointestinal tracts (gut) of humans or animals, there are microbes existing in the environment that can, in some cases, cause disease in humans (Prescott et al. 1993). Escherichia coli (E. coli) are a common intestinal bacteria found in large concentrations within warm-blooded animals. E. coli are commonly used as indicator organisms, indicating evidence of faecal material (WHO 2003). Salmonella spp. are a group of human pathogens that can infect the gastrointestinal tract of humans, causing diarrhoea (Prescott et al. 1993). Pseudomonas aeruginosa is an opportunistic pathogen that may cause infection through skin lesions (WHO 2003). Enterococcus spp. is an anaerobic bacterial genus that is a commensal inhabitant of the human intestine. It is reported to provide a higher correlation with many of the human pathogens found in wastewater than faecal coliforms (Jin et al. 2004).

Assessment of the health risk of harvested rainwater requires consideration of whether a hazard to human health is present and whether the dose of the hazardous substance is sufficient to cause illness.

6.3   Persistence of Waterborne Pathogens and Growth in Water

While typical waterborne pathogens are able to persist in drinking water, most do not grow or proliferate in water. Microbial quality of drinking water is commonly measured by testing for Escherichia coli (E. coli) or alternatively thermotolerant coliforms (often referred to as faecal coliforms) as indicators of possible contamination. Microorganisms such as E. coli can accumulate in sediments and are mobilised when water flow increases. After leaving the body of their host, most pathogens gradually lose viability and the ability to infect (Prescott et al. 1993). The rate of decay is usually exponential, and a pathogen will become undetectable after a certain period. Pathogens with low persistence must rapidly find new hosts. Persistence is affected by several factors, of which temperature is the most important. Decay is usually faster at higher temperatures, hence, the term 'thermal inactivation'. For pathogen-contaminated water to cause illness in humans, the pathogens must have an available route of infection and must overcome the defence barriers of the human body (stomach acidity, competition by natural gut flora and immunological responses, including acquired immunity). Routes of infection may include inhalation or ingestion. Successful infection by the pathogen is ultimately dependent on the pathogen dose (Prescott et al. 1993).

6.4   Public Health Risks Associated with rwh

About three million Australians use roof-harvested rainwater from tanks for drinking in urban and rural regions (ABS 1994). The *Australian Guidelines for Water Recycling: Managing Health and Environmental Risks 2008,* provide an assessment of the likely risks associated with the intended use and exposures for alternative water supplies. Fuller et al. (1981), Mobbs (1998) and Cunliffe (1998) found that the quality of rainwater was often adequate for potable uses provided the rainwater tank and roof catchment were subject to adequate maintenance (Coombes et al. 2002). Studies from both Ireland (O'Hogain et al. 2012) and New Zealand (Simmons et al. 2001) found that the rwh supplies sometimes exceeded drinking water guidelines for lead and microbial indicator organisms. Cunliffe (1998) stated that the probable source of indicator bacteria detected in rainwater tanks was excreta from small animals and birds. Of the epidemiological studies conducted on rwh to date, two have compared the rates of gastroenteritis in young children. A study of crypto-sporidiosis notifications in South Australia found a significantly reduced risk of cryptosporidiosis associated with tank rainwater compared to public mains water (Weinstein et al. 1993). A cohort study of one thousand and sixteen four to six-year-old children, who were regular consumers of tank rainwater, concluded that they were at no greater risk of gastroenteritis than those who drank treated public mains water (Heyworth et al. 2006). The majority of rainwater tanks in this study were galvanised iron (59%) with 43% of tanks greater than 10 years old. Only 8% of the tanks had first-flush devices, and sludge was never removed in 42% of the tanks (Heyworth et al. 2006). A pooled epidemiological study of 13 case studies, which quantified the risk of gastrointestinal disease from rainwater consumption, concluded that there was no significant difference in risk comparing rainwater to improved water supplies (Dean and Hunter 2012). A double-blinded randomised controlled trial among 300 households in Adelaide, South Australia, concluded that there were no appreciable differences in health outcomes from drinking untreated or treated rainwater (Rodrigo et al. 2011). Ahmed et al. (2011) carried out a review of available research reporting the microbial quality of rwh. This review suggested that the quality of rwh is strongly influenced by the season and the number of preceding dry days (Kus et al. 2010; Lye 2009). Several of the case studies reviewed suggested links between gastroenteritis and consumption of untreated rainwater (Brodribb et al. 1995; Franklin et al. 2009; Murrell and Stewart 1983). These reported outbreaks tended to involve small numbers of individuals, and the reported illnesses were often related to communal rwh systems.

Coombes (2015) hypothesised that rwh contains an inherent water treatment process ('treatment train') consisting of flocculation, settlement, sorption and bioreaction and that stored rainwater quality improves as metal and chemical contaminants settle to form sludge. This study highlighted the importance of first-flush devices to remove 11–94% of dissolved solids and 62–97% of suspended solids from the first 0.25 mm of rainfall depth on the roof surface. This is achieved by designing a first flush device to divert this initial volume away from the rwh system. So for example, lets say the effective catchment area of a roof surface is $54m^2$. The volume of rainwater diverted would be 54 x (0.25/1000) = $0.0135m^3$ (or 13.5 litres). So if we had a pipe diameter of 0.16m (pipe radius 0.08m), the length of the first flush pipe needed would be calculated as the volume of diverted water / cross sectional area of the pipe = $(0.0135/(3.14 \times (0.08^2)))$ = 0.67m. Simarly if we wanted to divert the first 0.5mm of rainfall depth we would need to capture a volume of 27 litres with a first flush pipe length of 1.35m. Martin et al. (2010) assessed the microbial properties of rwh at two study sites at Newcastle on the east coast of Australia. They concluded that rainfall events contributed to the bacterial load in rainwater storage systems, but that processes within the rainwater storage ensured that these incoming loads were not sustained. Spinks et al. (2005) and Spinks (2007) concluded that biofilms that formed on rwh tank walls and at the base of the sludge layer act to improve water quality. Spinks (2007) concludes that the settlement of particulate matter to the bottom of rainwater tanks is probably the single most beneficial process within rainwater storage. The quality of rainwater supplies was not compromised by the accumulation of sludge in tanks. This confirms the observation by Coombes (2002) that it is preferable to avoid disturbing rainwater storages.

Further studies (Morrow et al. 2007, 2010) concluded that the majority of the rainwater-harvesting systems in national investigations were compliant with the chemical and metal values in Australian Drinking Water Guidelines. Given the variability of rwh quality throughout the system, the most reliable sampling location is point the of use. Evans et al. (2009) used pioneering medical science techniques including polymerase chain reaction (PCR) methods (not experimental real-time analysis) to extract deoxyribonucleic acid (ribonucleic acid) of microbes in over 40 rainwater harvesting systems over a 3-year period. Each sample was also subjected to a comprehensive range of microbial, medical and biochemical tests to confirm the results of the PCR analysis. The research found that bacteria of faecal origin were rare and not abundant or persistent in rainwater harvesting systems. This research discovered that rainwater storages act as balanced ecosystems in a similar fashion to environmental systems that improve water quality (Evans et al. 2009).

6.5 Rainwater and Hot Water Systems

The use of harvested rainwater in hot water systems is also not common in many developed countries, the notable exception Australia. Figure 6.1 shows the average water consumption results from an 18-month study carried out by the authors to investigate domestic household water consumption in Ireland. It shows that toilet use represented on average 22% of the total household water use. Hot water consumption represented an additional 20% of total usage. Hot water systems therefore represent a logical extension for the use of rwh to reduce mains water demand. However, concerns over possible increase risks to public health from rwh fed hot water systems has limited their potential usage.

Arguably, the most significant health risk in hot water systems comes from the respiratory pathogen Legionella pneumophila. L. pneumophila is the agent of Legionnaires disease, an acute form of pneumonia, which most commonly infects the respiratory tract of immunocompromised individuals. The most likely route of infection of the respiratory tract occurs when the bacteria is entrapped in aerosols and inhaled. The ingestion of L. pneumophila is harmless as they are unable to cope with the stresses of the gastrointestinal tract. There may be a potential health risk from showering in hot water if the water supply contains L. pneumophila and the hot water is maintained below 60 °C, as contaminated aerosols may be produced. However, this risk is equally applicable to mains water users. Thermal inactivation can be used as an effective method to inactivate Legionella bacteria. The degree of inactivation is dependent on the temperature and the length of time the bacteria are exposed to that temperature. The thermal inactivation of Legionella bacteria starts around 50 °C but is more rapid at higher temperatures. At 60 °C, 90% of L.

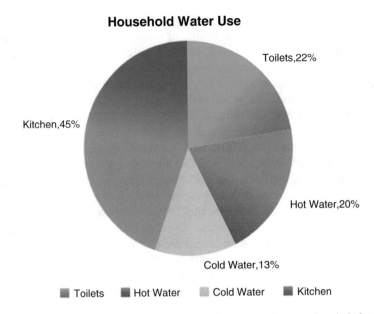

Fig. 6.1 Typical household water consumption rates from a study by the authors in Ireland

pneumophila will be inactive after 3.2 min of exposure (average value) (Makin 2014). Where the water contains 100,000 CFU/l Legionella, the bacteria need to be held at 60 °C for approximately 10 min to reduce numbers to below UK's Health and Safety Executive (HSE) action level of 100 CFU/l. Hot water storage cylinders that maintain a temperature of 60 °C throughout the whole storage vessel for a period of 1 h daily should achieve satisfactory control of Legionella bacteria, in line with the recommendations in UK's HSE code of practice (HSE 2013). A study by Borella et al. (2004) investigated Legionella spp. and Pseudomonas spp. contamination of hot water systems in Italy. Legionella spp. and Pseudomonas spp. were detected in 22.6 and 38.4% of samples, respectively. The study found that system and building characteristics were the main predictors for Legionella. Legionella contamination was associated with a centralised heater, distance from the heating point of >10 m and a water plant that is >10 years old. Legionella presence was not affected by the origin of water (Borella et al. 2004). A study by Kruse et al. (2015) in Germany analysed water samples from 718 buildings for Legionella spp. The study concluded that the most important risk factor for contamination with Legionella spp. was the temperature of the circulated hot water.

6.6 Heat Inactivation Rates for Waterborne Disease

While extensive research has been undertaken in the food industry to determine the heat inactivation rates for pathogens, studies for thermal inactivation in a freshwater medium are more recent and more specialised (Spinks et al. 2005). Spinks et al. (2006) carried out thermal inactivation analyses on eight species of non-spore form- ing bacteria in a water medium at temperatures of 55–65 °C, and susceptibilities to heat stress were compared using D-values. The D-value for this study was defined as the time required to reduce a bacterial population by 90% or 1 log reduction. The results suggested that the temperature range from 55 to 65 °C was critical for the effective elimination of enteric/pathogenic bacterial components and supported the thesis that hot water systems should operate at a minimum of 60 °C. The study also recommended that future rainwater harvesting investigations should focus on the microbial ecology of rainwater treatment trains and stored water to determine the types of organisms likely to exist in these systems. Ahmed et al. (2011) also con- cluded that any microbial assessment should involve the analysis of rwh for actual pathogenic species, not just the common faecal indicator bacteria. Coombes et al. (2006) and Evans et al. (2006) further highlighted that E. coli has a large number of non-faecal environmental strains that are prevalent in natural waters (such as rain- water). Later research by Luo et al. (2011) confirmed this key issue – that the isola- tion of E. coli in rainwater may not indicate faecal contamination.

Thermal Inactivation Results

Laboratory experiments were conducted by the authors using a variety of water- related bacteria to determine the time required to reduce a bacterial population by 90% at a given temperature (McCarton and O'Hogain 2017). Analysis was carried out to assess the thermal destruction rates of sterile water samples spiked with known quantities of bacteria. The temperatures chosen were 55 and 60 °C as they mimic the conditions typically found in a domestic hot water system. Aliquots of the spiked water samples were taken at times 0, 5, 10, 15 and 20 min and processed for a concentration of bacteria. The bacteria chosen for the experiments included E. coli, Enterococcus fecalis, P. aeruginosa and Salmonella spp. The results of this study showed that after 5 min of exposure at 60 and 55 °C, respectively, Salmonella, Pseudomonas aeruginosa and total viable count at 22 and 37 °C concentrations were reduced to zero. Irish standards require hot water systems to be maintained at tem- peratures at or above 60 °C. The conclusion from this pilot study were that thermal inactivation in rwh fed hot water systems at a minimum of 60 °C, combined with the treatment train inherent within rwh systems, is likely to deliver water quality com- parable to mains hot water. From the literature, the main risk to public health from hot water systems is not the water source but rather the operation, maintenance, age, location and temperature of the hot water system. Hot water systems supplied with harvested rainwater do not present an increased risk to health over hot water sys- tems fed with mains water.

Chapter 7
Answering the Demand Versus Supply Question

© Springer Nature Switzerland AG 2021
L. McCarton et al., *The Worth of Water*,
https://doi.org/10.1007/978-3-030-50605-6_7

7.1 Introduction

Rainwater – an untapped resource
A breakdown of per capita consumption (pcc) rates for a typical household is shown. In municipal supply systems currently all water supply from mains is treated to potable (drinking) water standard. However, when we review the actual household water usage, over 50% of a household's water demand may be for non-potable water uses, such as garden use, toilet flushing and washing machines. Therefore, it makes sense to develop the science and engineering to utilize this potential untapped resource to augment domestic demand from on-site rainwater harvesting. Figure 7.1 shows typical household water consumption rates.

Typical Household Water Use

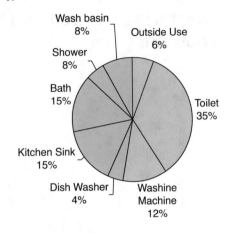

Fig. 7.1 Typical Household water consumption rates

7.2 Water Flows Within the rwh Process

Rainwater harvesting for the purpose of this discussion shall be confined to rain collected from roof surfaces only.

A typical rainwater harvesting system consists of a number of basic components:

- catchment surface;
- collection area;
- conveyance system;
- filter device(s);
- storage tank; and
- delivery system to transport the rainwater to the point of use.

The water flows within the rwh process are shown in Fig. 7.2.

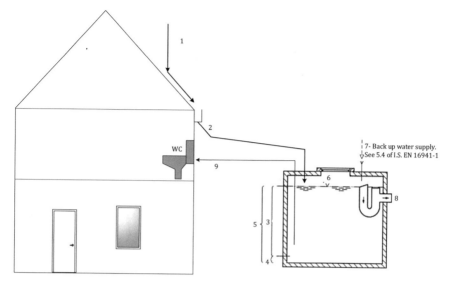

Key

1   Rainwater collection area
2   Conveyance system
3   Volume of usable capacity of storage tank
4   Minimum water level of storage tank
5   Nominal capacity of storage tank
6   Maximum water level of storage tank
7   Back up water supply. See 5.4 of I.S. EN 16941-1
8   Overflow returned to rain water disposal system
9   Non-potable water supply

Fig. 7.2   Schematic showing rainwater harvesting water flows

The water flows shown in Fig. 7.2 are further described as follows:

- The harvested rainfall (1) is collected from the collection area and conveyed (2) to the storage tank;
- The daily non-potable water demand (9) is supplied by either the harvested rainwater from the storage tank or a backup water supply (7), or a combination of both if required;
- Any harvested rainwater in excess of the nominal capacity (5) of the storage tank overflows to a rainwater disposal system (8);

7.3   Determining if a rwh System Is Feasible for Your Site

Figure 7.3 shows the design methodology used to establish if rwh is feasible for your site. The demand profile should be established. Then the Supply profile is established by calculating how much rainwater can be harvested. This is dependent on the rainfall, catchment characteristics and system losses. The amount of harvestable rainwater is then compared to the demand for this harvested rainwater. This allows the designer to establish if it is feasible to proceed with this supply option. Table 7.1 summarises the steps in this design methodology.

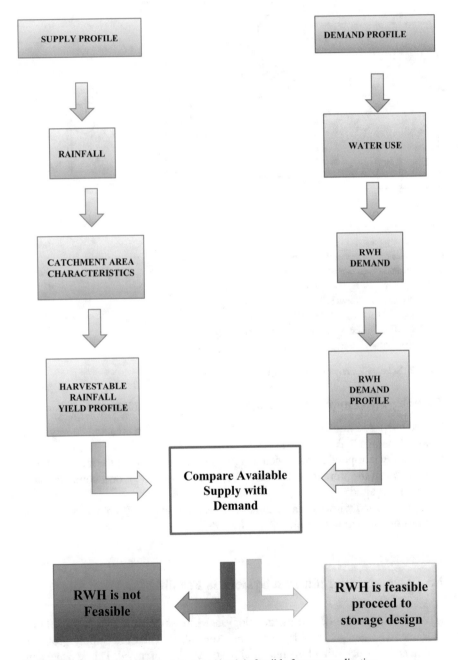

Fig. 7.3 Design Methodology to determine if rwh is feasible for your application

Table 7.1   RWH demand vs supply design methodology

rwh Design Methodology – Stage 1 Hydraulic Efficiency		
STEP 1	Obtain rainfall records and establish the rainfall pattern for the site location Plot the monthly rainfall and note any seasonal variations.	Average Annual Rainfall Average Monthly Rainfall Daily Rainfall (if available)
STEP 2	Calculate the Harvestable Rainwater Yield ($Y_{R,t}$) Plot the monthly harvestable rainwater yield versus the monthly rainfall	$Y_{R,a\,(ANNUAL)}$ $Y_{R,m\,(MONTHLY)}$ $Y_{R,d\,(DAILY)}$
STEP 3	Calculate the Demand Profile for the site ($D_{N,t}$)	$D_{N,a\,(ANNUAL)}$ $D_{N,m\,(MONTHLY)}$ $D_{N,d\,(DAILY)}$
STEP 4	Summarise the results from stage 1 in a table Compare available rwh supply versus demand, and calculate the Annual and Monthy Supply Coefficient (S).	$S = Y_{R,t}/D_{N,t}$
rwh Design Methodology – Stage 2 rwh Storage Tank Sizing		
	Calculate optimum rwh storage volume required using one of the following; (i) Basic method (ii) Tabular method (iii) Detailed daily storage model	

Stage 1 involves establishing the hydraulic efficiency of the rwh system. This is the percentage of demand that the rwh system can supply. Stage 2 involves selecting the appropriate rwh storage solution according to one of three design methods. This methodology is further described in the following sections.

7.4   Stage 1 Hydraulic Efficiency of the rwh System

STEP 1: ESTABLISH THE RAINFALL PROFILE FOR THE SITE LOCATION

What is the rainfall pattern for my site?

The starting point in any rainwater harvesting design process is to review the long-term rainfall records measured at a location representative of the rainfall at the site. Most rainfall gauging stations itemise the depth of rainfall falling over a 24-hour period. Daily records are usually then collated to give monthly rainfall depths. These are often represented as average values from rainfall records recorded over a number of years. It is important to review seasonal and yearly rainfall patterns in order to predict how the rainwater harvesting system is likely to meet demand and in order to estimate any required storage and/or back up water system required.

STEP 2: CALCULATE THE HARVESTABLE RAINWATER YIELD ($Y_{R,t}$)

What volume of rainfall can I potentially harvest from my site?

Having established the rainfall pattern for the site location, the next step is to review the characteristics of the collection surface and calculate the maximum volume of rainwater which can be harvested from the collection area. This is termed the "Harvestable Rainwater Yield ($Y_{R,t}$).This can be expressed as annual yield ($Y_{R,a}$), monthly yield ($Y_{R,m}$) or daily yield ($Y_{R,d}$) depending on the available rainfall records and the level of accuracy required.

Harvestable Rainfall Yield, $Y_{R,t}$

($Y_{R,t}$) is the rainwater yield per time step t, expressed in litres per timestep and should be calculated using Eq. 7.1

$$Y_{R,t} = A \times h \times e \times \eta \qquad (7.1)$$

where

A is the horizontal projection of the collection area (m²) which is to be drained to the rwh system
h is the total rainfall (mm) for a chosen timestep t (daily, monthly or yearly);
e is the surface yield coefficient
η is the hydraulic treatment efficiency coefficient

Note The timestep chosen for the rainfall depth (h) will determine the Y_R value. For example, if the average rainfall depth for the year is used in the formula, the corresponding annual yield ($Y_{R,a}$) will be calculated. Similarly, if the average rainfall depth for the month is used the formula will calculate the corresponding monthly yield ($Y_{R,m}$).

Collection Area (m²)

The collection area (A) is the horizontally projected area of the collection surface (roof) which can be drained to the rwh system. It is measured in m². It should only include those surfaces from which rainwater is collected and conveyed to the storage tank.

Rainfall Depth (h)

The annual average rainfall depth (mm) should be representative of the site location and can usually be obtained from the nearest national meteorological monitoring station. If a monthly rainfall profile is available for the site this can show potential seasonal variabilities in potential supply.

Surface Yield Coefficient (e)

Collection surfaces composed of different materials have different characteristics regarding the drainage of rainwater. The volume of harvested rainwater is influenced by the surface yield coefficient, e. This is defined as the ratio of the volume of water which runs off a surface to the volume of rainfall which falls on the surface, i.e. a measure of the volume of rainfall which can be captured and stored.

$$e = (\textbf{VOLUME OF RUNOFF} / \textbf{VOLUME OF RAINFALL}) \qquad (7.2)$$

A surface yield coefficient of 0.9, indicates that approximately 90% of the rainfall which falls on a catchment surface can be captured.

The surface yield coefficient is dependent on the roof collection surface. Table 7.2 summarises the surface yield coefficient for different collection surfaces.

Table 7.2 Surface Yield Coefficient (adapted from BS EN 16941–1:2018)

Collection Surface	Surface Yield Coefficient (e)
Pitched smooth surface roof (e. g. metal, glass, slate, glazed tiles, solar panels)	0.9
Pitched rough surface roof (e. g. concrete tiles)	0.8
Flat roof without gravel	0.8
Flat roof, with gravel	0.7
Green roof, intensive (e.g. garden)	0.3
Green roof, extensive	0.5
Sealed areas, (e.g. asphalt)	0.8
Non-sealed areas (e.g. cobble stone)	0.5

Hydraulic treatment efficiency coefficient (η) Protecting the quality of the harvested rainwater requires that roof debris and dirt be excluded from entering the storage system. Introducing a filtration system before the collected rainwater enters the storage tank prevents debris accumulating in the tank. A filtration system should include a filter that has the following characteristics;

- water and weather resistant
- removable and readily accessible for maintenance purposes
- has an efficiency of at least 90%
- passes a maximum particle size of <1.25 mm.

The hydraulic treatment efficiency coefficient is the ratio of the outcoming flow of filtered water to incoming flow of the collected rainwater from the roof surface. The manufacturer will normally supply this coefficient if requested. This can also include for water lost during a first flush device. Typical hydraulic treatment efficiency coefficients are in the range 0.75–0.9. If this value is not available from the manufacturer 0.9 is normally assumed.

Harvestable Rainwater Yield

A useful rule of thumb to remember is that excluding system losses:

1 square metre (m^2) of roof catchment × 1 millimetre (mm/yr) of rainfall = 1 litre of harvested rainwater/yr. We can then apply system losses for surface yield coefficient (e = 0.9) and hydraulic treatment efficiency ($\eta = 0.9$)

$$Y_{R,t} = A \times h \times e \times \eta \qquad (7.3)$$

For example: 1 (m^2) × (1/1000) m × 0.9 × 0.9 = 0.00081 m^3/yr = 0.81 litres/yr

STEP 3: CALCULATE THE DEMAND PROFILE

What quantity of water is needed and when?

On-site collection and use of rainwater cover a variety of applications such as toilet flushing, laundry, irrigation and garden use, agriculture, climate control of buildings, cleaning etc. The typical hierarchy is to prioritise harvested rainwater to replace non-potable water demand onsite. Non-potable water is defined as water which does not meet drinking water quality in accordance with Council Directive 98/83/EC. If there is excess harvestable rainwater supply over non-potable demand, then consideration can be given to further treating the harvested rainwater to meet additional water demand onsite. With appropriate treatment rwh systems can be designed to supply potable water standards.

Determination of the non-potable water demand per day ($D_{N,d}$)

The total daily non-potable water demand per household, $D_{N,d}$ (l/d) is estimated based on the forecasted uses, their frequency and their seasonality. The demand can vary substantially according to the region, climate and type of building. Occupancy levels and socio-economic status also influence the demand.

The non-potable water demand (litres/day) should be calculated using Eq. 7.4:

$$D_{N,d} = D_{p,d} \times n \qquad (7.4)$$

where

$D_{p,d}$ is the daily per-person non-potable water demand (l/(p × d));
n is the number of persons in the connected building(s).

Note In Europe the value of $D_{p,d}$ is taken as 60 litres per person per day (l/(p × d)) which comprises of 50 l/(p × d) for WC flushing and clothes washing and 10 l/(p × d) for miscellaneous use.

Determination of Monthly Non-Potable Water Demand ($D_{N,m}$)

The total monthly non-potable water demand ($D_{N,m}$) expressed in litres per month (l/mth) should be calculated using Eq. 7.5:

$$D_{N,m} = D_{N,d} \times number\ of\ days\ in\ the\ month \qquad (7.5)$$

Where:

$D_{N,d}$ is the total daily non-potable water demand per household calculated using Eq. 7.4.

Determination of Total Annual Non-Potable Water Demand ($D_{N,a}$)

The total annual non-potable water demand ($D_{N,a}$) expressed in litres per annum (l/a) should be calculated using Eq. 7.6:

$$D_{N,a} = D_{N,d} \times 365 \qquad (7.6)$$

where

$D_{N,d}$ is the total daily non-potable water demand per household calculated using Eq. 7.4.

STEP 4: COMPARE AVAILABLE HARVESTABLE RAINWATER YIELD VERSUS DEMAND

The amount of water which can be provided from rwh will be limited by the rainfall profile, collection area connected to rainwater system, size of storage tank and required demand. If monthly rainfall and demand figures have been established in Steps 1 and 2, these should be plotted to give a visual representation of the potential performance of the rwh system within different seasons. This will allow the designer to make an informed decision as to the feasibility of proceeding to detailed design stage.

Stage 1 Design Summary

It is useful to summarise the design to date using the tabular structure shown in Table 7.3.

Column 1 is the months
Column 2 summarise the average monthly rainfall depths (mm)
Column 3 shows the monthly harvestable rainfall yield (m³)
Column 4 shows the monthly demand (m³)
Column 5 calculates the deficit or surplus for the month, equal to column 3 minus column 4. **A negative sign indicates that demand exceeds rwh supply for this month (deficit).** Demand will have to be met by either stored rainwater from a previous month (if possible) or a back-up water supply. **A positive sign indicates that rwh supply exceeds demand for this month (surplus)** and this surplus could potentially be stored for reuse in later months.
Column 6 shows the **supply coefficient which presents the monthly rwh supply as a percentage of demand.** We can plot a graph of monthly harvestable rainfall yield (column 3) versus monthly demand (column 4). We can also plot monthly harvestable rainfall yield (column 3) versus monthly demand (column 4) and potential storage volume (column 5). This shows graphically the predicted potential performance of the rwh system. It will also allow us to determine if rwh is feasible for this particular site.

Table 7.3 Tabular Representation of Stage 1 Design Summary

Site Location:					
1	2	3	4	5	6
Month	Rainfall	Harvestable rainfall yield	Demand	Deficit or Surplus	Supply coefficient
		$Y_{R,m}$	$D_{N,m}$	$Y_{R,m}-D_{N,m}$	$Y_{R,m}/D_{N,m} \times 100$
	(mm)	(m³)	(m³)	(m³)	%
Jan					
Feb					
March					
April					
May					
June					
July					
Aug					
Sept					
Oct					
Nov					
Dec					
Total					
Avg. monthly rainfall					

Annual rwh Supply Coefficient ($S=Y_{R,t}/D_{N,t}$)

Using the annual rainwater yield and annual non-potable demand figures respectively the annual rwh supply coefficient (S) can be established using Eq. 7.7:

$$\text{Annual Supply Coefficient}, S\left(\%\right) = \frac{Y_{R,a}}{D_{N,a}} \times 100 \qquad (7.7)$$

Table 7.4 shows different interpretations of S.

Table 7.4 Making sense of the annual rwh supply coefficient

Annual rwh Supply Coefficient	Decision
S > 100%	This indicates that on an annual basis, the harvestable rainwater yield is greater than the annual non-potable water demand for the site. Excess harvestable rainwater could potentially be stored and treated to potable standard and used to meet additional demand on site.
S < 100%.	This indicates that on an annual basis, the harvestable rainwater yield is less than than the annual non-potable water demand for the site. The harvestable rainwater yield will have to be supplemented by an alternative water source to meet non-potable water demand.
S = 100%	This indicates that on an annual basis, the harvestable rainwater yield is equivalent to the annual non-potable water demand for the site.

See worked examples 7.1 and 7.2.

Stage 1 Feasibility

You will now have compiled an assessment of the potential hydraulic efficiency for rainwater harvesting at the site location. The designer can then assess whether it is feasible to proceed to Stage 2.

7.5 Stage 2 Rainwater Storage Tank Sizing

What volume of rainwater storage is required?

If the results from the hydraulic efficiency calculations show that rainwater harvesting can meet the required demand, any excess can be stored. The designer then proceeds to determine the optimum rwh storage tank size. Rainwater storage tank sizing can be determined using either a basic numerical method, an intermediate tabular method or a detailed method which establishes a daily storage model of the system.

(i) A Basic Numerical Method can be used for rwh projects with regular demand and a constant annual rainwater profile.
(ii) The Tabular Method is appropriate for locations with variable demand patterns and/or variable annual rainfall profiles.
(iii) A Detailed Method can be utilised for complex non-domestic applications where the demand is irregular.

BASIC APPROACH

To apply the basic approach to size rainwater harvesting systems for non-potable domestic use, the nominal storage capacity should be the lesser of 5% of the annual rainwater yield ($Y_{R,a}$) or 5% of the total annual non-potable water demand ($D_{N,a}$).

Note: This method assumes an 18-day dry period ($18/365 = 5\%$)

EXAMPLE

Taking a house located in Dublin, Ireland with a 4-person household with a roof area of 70 m² located in a region with a total annual rainfall of 557 mm (based on an annual average of 10 years rainfall data), with a surface yield coefficient of 0.9 and hydraulic treatment efficiency coefficient of 0.9.

$$Y_{R,a} = A \times h \times e \times \eta$$
$$Y_{R,a} = 70 \times 557 \times 0.9 \times 0.9$$
$$Y_{R,a} = 31,581.90 l/a \quad \text{or} \quad 31.58 \, m^3/a$$
$$5\% Y_{R,a} = 1.58 \, m^3$$

Assumed daily per-person non-potable water demand, $D_{p,d} = 60$ l/(p × d)
Number of persons in house = 4
The total daily non-potable water demand:

$$D_{N,d} = 60 \times 4 = 240 \, \text{l/d} \quad \text{or} \quad 0.24 \, \text{m}^3 \text{/d}$$

The total annual non-potable water demand:

$$D_{Na} = 240 \times 365 = 87{,}600 \, \text{l/a} \quad \text{or} \quad \left(87.6 \, \text{m}^3 \text{/a} \right)$$

$$5\% D_{N,a} = 4.38 \, \text{m}^3$$

The nominal storage capacity should be the lesser of 5% of the annual rain-water yield ($Y_{R,a}$) or 5% of the total annual non-potable water demand (D_{Na}).

Therefore, in this case the nominal storage capacity required would be 1.58 m³. This example illustrates a site where the harvestable rainfall yield is the governing design storage parameter.

TABULAR METHOD

In low rainfall areas or areas where rainfall is of uneven distribution (i.e. distinct wet/dry seasons) or where demand is not constant through the year, a tabular method may be more appropriate to estimate the nominal capacity of the rainwater storage required. During some months of the year there may be an excess of rainwater. If there is sufficient excess to balance periods where demand exceeds rainfall supply, then it may be possible to design a rwh storage system which meets demand through all seasons. For these situations it may be more appropriate to utilise the intermediate tabular approach to determine the rwh storage capacity required.

Finding the minimum size of tank using the tabular method

Table 7.5   Tabular Method

1	2	3	4	5	6	7	8	9
		Harvestable rainfall yield $Y_{R,m}$	Demand $D_{N,m}$	Deficit or Surplus $Y_{R,m}-D_{N,m}$	Supply Coefficient $Y_{R,m}/D_{N,m} \times$ 100	Cumulative Harvestable Rainfall Yield $\sum Y_{R,m}$	Cumulative Demand $\sum D_{R,m}$	Potential Storage Volume $\sum Y_{R,m}-$ $\sum D_{R,m} \times$ 100
	Rainfall							
Month	(mm)	(m³)	(m³)	(m³)	(m³)	(m³)	(m³)	(%)
January								
February								
March								
April								
May								
June								
July								
August								
September								
October								
November								
December								
Total								

The basis of the tabular method is to construct a table as shown in Table 7.5.

Column 1 – Months
Column 2 – Average monthly rainfall depths from available records
Column 3 – Harvestable Monthly Rainfall Yield calculated from Eq. 7.3
Column 4 – Monthly Demand
Column 5 – Monthly Deficit or Surplus (column 3 minus column 4)
Column 6 – Monthly Supply Coefficient (%)
Column 7 – Cumulative Harvestable Rainfall Yield
Column 8 – Cumulative Demand
Column 9 – Potential Storage Volume (= column 7 – column 8)

DETAILED APPROACH

General

This approach should be used to accurately calculate the storage capacity required for all situations, in particular where:

– the demand is irregular (e.g. external use, non-residential use, tourism);
– the yield is uncertain (e.g. due to the use of green roofs, permeable pavements); and
– larger or complex rainwater harvesting systems are proposed.

Input data

The following information is required for carrying out the detailed approach:

– daily rainfall on the site, h_d (mm/d), for a minimum of a 5-year period (preferably recent years); and,
– total daily non-potable water demand, $D_{N,d}$ (l/d).

Simulation principle

The use of the detailed approach method shall consider the mass balance of the rain-water storage system. In this case there are two inputs (rwh supply and if appropriate, a back-up water supply) and two outputs (daily demand requirements and overflow). This method involves carrying out a daily assessment of the inputs and outputs over a defined period to determine the appropriate usable volume of a rwh storage tank. The parameters to be considered in this method are illustrated in Fig. 7.4.

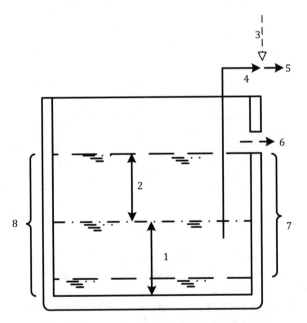

Fig. 7.4 Schematic of design parameters to be used in the detailed approach

Key	
1	$V_{r(d-1)}$ – Volume of rainwater in the storage tank from previous day (l)
2	$Y_{r,d}$ – Harvested rainwater entering the storage tank during current day, (l/d)
3	M – Back up water supply. See 5.4 of I.S. EN 16941–1, (l/d)
4	$S_{N,d}$ – Daily non-potable water abstracted from storage tank, (l/d)
5	$D_{N,d}$ – Total daily non-potable water demand, (l/d)
6	O_d – Overflow returned to rain water disposal system, (l/d)
7	V – Useable volume of storage tank, (l)
8	S – Nominal capacity of storage tank, (l)

Timesteps

The mass balance of the system, according to timesteps, shall be considered where:

- Timestep, d-1, refers to the previous day; and
- Timestep, d, refers to the current day.

What happens when rainfall enters the system during the current timestep (d) depends on the volume of water already in the storage tank from the previous day (d-1). This volume is denoted as $V_{r(d-1)}$. This is the starting point in the model. The volume of harvested rainwater entering the storage tank during the current timestep shall be denoted as $Y_{r.d}$.

- If $Y_{r.d} + V_{r(d-1)}$ is less than the useable volume of the storage tank (V), then the rainwater will be stored in the system. This water, which is stored in the tank, is then available for supply to the building(s).
- If $Y_{r.d} + V_{r(d-1)}$ is greater than V, then the rainwater will overflow from the system. The rainwater overflowing from the system during the current timestep, is termed O_d.

The rainwater system will operate according to one of the following:

- If the total daily non-potable water demand ($D_{N,d}$) is greater than the daily non-potable water abstracted from the storage tank ($S_{N,d}$), then a backup water supply (M) will be required to meet the non-potable water demand; or
- If $D_{N,d}$ is less than $S_{N,d}$ the rainwater system will meet 100% of the non-potable water demand and the volume in the storage tank will be reduced accordingly.

The operation of the rainwater system may be expressed using Eq. 7.8

$$S_{N,d} = \min \begin{cases} D_{N,d} \\ V_{r(d-1)} \end{cases} \tag{7.8}$$

The daily non-potable water abstracted from the storage tank ($S_{N,d}$) will be the lesser of the total daily non-potable water demand ($D_{N,d}$) in the current time period, or the volume of rainwater in the storage tank from the previous day ($V_{r(d-1)}$).

The predicted volume of harvested rainfall stored in any one day ($V_{r.d}$) is governed by two conditions:

1. If $Y_{r.d} + V_{r(d-1)}$ is greater than useable volume of the storage tank (V), then the excess rainwater will overflow from the system (O_d); and,
2. The useable volume of the storage tank (V) less the daily non-potable water abstracted from the storage tank ($S_{N,d}$).

The predicted volume of harvested rainfall stored in any one day ($V_{r.d}$) will be the lesser of that from the two conditions and may be expressed using Eq. 7.9.

$$V_{r.d} = \min \begin{cases} Y_{r.d} + V_{r(d-1)} - O_d \\ V - S_{N,d} \end{cases} \tag{7.9}$$

This now allows the designer to apply the detailed approach to develop a spreadsheet to model the daily performance of the rainwater harvesting system.

CAUTION — To apply the detailed approach, specialist expertise should be sought, as the detailed approach to rainwater harvesting design is site specific, i.e. it is applicable only to the site for which the design is being produced.

The detailed approach requires either hourly or daily rainfall data for the specific site covering at least a five-year period. Detailed information on water demand is also required and this should be acquired by monitoring water demand over an extended period of time, to accurately establish daily and monthly variations. Applications of any detailed design are, by definition, only applicable to the site for which it is being designed. These sites, for example, a school, a hospital, an airport terminal or a sports arena require design by a specialist who has access to the relevant rainfall data and the relevant data on water demand.

Worked Examples

To illustrate the design process, we will select two worked examples from different geographical regions to assess the feasibility of utilising rainwater harvesting to replace non-potable demand (toilet flushing) in a domestic house.

7.6   Worked Examples

Worked Example 7.1 STEP 1: ESTABLISH THE RAINFALL PROFILE FOR THE SITE LOCATION

What is the rainfall pattern for my site?

Table 7.6 shows the average monthly rainfall (mm) figures taken from 10 years records for a site. This is in a temperate climate region with evenly distributed rainfall over the 12 months.

Table 7.6 Average monthly rainfall depths (mm) for Dublin, Ireland

Jan	Feb	Mar	Apr	May	June	July	Aug	Sept	Oct	Nov	Dec	Total	Avg
64.86	32.60	46.61	26.72	30.20	33.98	49.23	55.06	39.03	51.93	74.16	52.11	557.49	46.46

Plot the monthly rainfall and note any seasonal variations (Fig. 7.5).

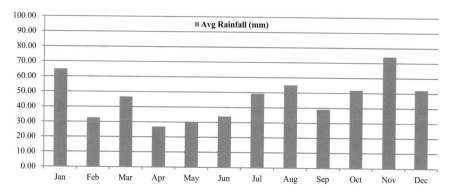

Fig. 7.5 Rainfall profile for site in Dublin, Ireland

This establishes a visual representation of the rainfall profile for the site. For this particular site it can be seen that rainfall is evident at this site across 12 months of the year. The total annual rainfall depth is 557.49 mm. The average monthly rainfall depth is 46.46 mm. There are 6 months when rainfall exceeded this average. Minimum rainfall occurs during April (26.72 mm) and maximum rainfall during November (74.16 mm). The ratio of maximum to minimum rainfall is 2.77. A ratio of less than 5 shows that the rainfall is relatively evenly distributed throughout the year.

STEP 2: CALCULATE THE HARVESTABLE RAINWATER YIELD ($Y_{R,t}$)
What volume of rainfall can I potentially harvest from my site?
Taking a tiled roof on a residential house with 70m² roof area draining to a single downpipe, the householder would like to harvest the rainwater for use within the house for toilet flushing. From Table 7.2 the surface yield coefficient (e) is 0.9. It is proposed to fit a downpipe filter with a hydraulic treatment efficiency coefficient of 0.9.

$$Y_{R,t} = A \times h \times e \times \eta$$

Y_{Ra} = 70 (m²) × 557.49 (mm/yr) × 0.9 × 0.9 = 31,609.68 litres per annum (31.61 m³ per annum)

For this site we also have the monthly rainfall depths available. Therefore, we can repeat the above calculation using monthly rainfall depths to establish the harvestable rainwater yield per month.

We can plot a profile of this analysis. This presents a useful graphical representation of any seasonal variation. For our site in Dublin, Fig. 7.6 shows that the harvestable rainfall yield is relatively consistent throughout the year.

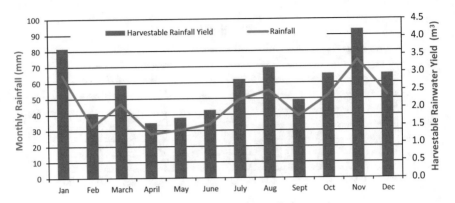

Fig. 7.6 Average monthly rainfall (mm) and harvestable rainfall yield for Dublin site

STEP 3: CALCULATE THE DEMAND PROFILE
What quantity of water is needed and when?

We can now establish the demand which we propose to supply from the harvested rainwater. For this example, we will assume a non-potable water demand per person of 60 l/(p × d) which comprises of 50 l/(p × d) for WC flushing and 10 l/(p × d) for miscellaneous external garden use. We will assume 4 persons in the house.

The non-potable water demand can be calculated using Eq. 7.4

$$D_{N,d} = 60 \times 4 = 240\,l/d\left(0.24\mathrm{m^3/d}\right)$$

The total annual non-potable water demand ($D_{N,a}$) expressed in litres per annum (l/a) can be calculated as follows:

$$D_{N,a} = 240 \times 365 = 87,000\,l/a\left(87.6\mathrm{m^3/a}\right)$$

Similarly, we can calculate the monthly non-potable water demand, $D_{N,m}$ is equal to 7.3m³/month. It is assumed that this demand is constant throughout the year.

STEP 4: COMPARE AVAILABLE HARVESTABLE RAINWATER YIELD VERSUS DEMAND

We can summarise the design calculations in the following table:

Table 7.7 Tabular Method Results

Site Location: 1	Dublin, Ireland 2	3	4	5	6
Month	Rainfall	Harvestable rainfall yield	Demand	Deficit or surplus	Supply coefficient
		$Y_{R,m}$	$D_{N,m}$	$Y_{R,m}-D_{N,m}$	$Y_{R,m}/D_{N,m} \times 100$
	(mm)	(m³)	(m³)	(m³)	%
Jan	64.86	3.68	7.30	−3.62	50%
Feb	32.60	1.85	7.30	−5.45	25%
March	46.61	2.64	7.30	−4.65	36%
April	27.72	1.57	7.30	−5.72	22%
May	30.20	1.71	7.30	−5.58	23%
June	33.98	1.93	7.30	−5.37	26%
July	49.23	2.79	7.30	−4.50	38%
Aug	55.06	3.12	7.30	−4.17	43%
Sept	39.03	2.21	7.30	−5.08	30%
Oct	51.93	2.94	7.30	−4.35	40%
Nov	74.16	4.20	7.30	−3.09	58%
Dec	52.11	2.95	7.30	−4.34	40%
Total	**557.49**	**31.61**	**87.55**	**−55.94**	**36%**
Avg.	46.46				

Note: Column 5, a negative value indicates the volume of water required from a backup supply to meet demand, a positive value indicated potential excess rainwater which could be stored

Table 7.7 presents a prediction of how a rainwater harvesting facility may function at our site. We can see that demand exceeds supply for all months. Column 5 shows either the volume of additional water which is required to meet this monthly demand (indicated by a **negative value**) or the potential harvested rainwater which could be stored (indicated by a **positive value**). Column 6 shows the monthly supply coefficient, which is the percentage of demand which can be supplied from harvested rainwater. This ranges from a low of 22% in April to a max of 58% in November. The annual supply coefficient is 36%. This shows that over a 12-month period harvested rainwater can supply an average of 36% of non-potable demand for this household. The remaining demand will have to be met from a back-up water supply, typically from a municipal mains water supply.

We can also plot the harvestable rainfall yield versus monthly demand (Fig. 7.7).

This visual representation shows clearly the demand and supply profile. Demand is constant throughout the year at 7.3m³ per month. Supply (harvested rainfall yield) varies but is constantly less than demand for all months.

Finally, we can plot a comparison of harvestable rainfall yield with monthly demand and the corresponding deficit or surplus (Fig. 7.8).

This shows visually the performance of the rwh system at this site. It can be seen that column 5 is negative for all months indicating that there is a deficit for all months.

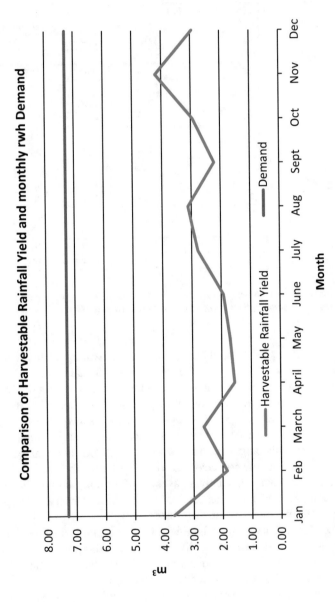

Fig. 7.7 Comparison of harvestable rainfall yield and monthly demand for example 1

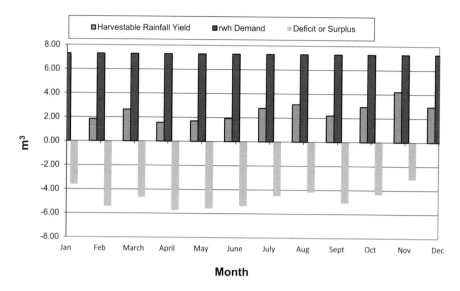

Fig. 7.8 Comparison of harvestable rainfall yield, monthly demand and potential storage volume

The overall conclusion as to whether rwh is hydraulically feasible at this location depends on;

- Is there a reliable back up supply to meet the non-potable water demand which cannot be met by harvested rainwater.
- Is the reduction in mains water savings a driver, if so then rwh at this location could reduce mains consumption by 36% annually.

This example is for a single household. However, it also illustrates for a local water supply authority the potential to reduce mains water demand across the network by up to 36% per household. If rwh formed a core part of a diverse water supply and management strategy this could potentially free up spare capacity across the network.

Conclusion: Rainwater harvesting is hydraulically feasible at this site, proceed to Design Stage 2.

STAGE 2 STORAGE DESIGN

(i) Basic Storage Design

This method recommends estimating the storage required to supply 5% of the annual rainwater yield or 5% of the annual non-potable water demand. The lesser value from these two equations is taken as the optimum tank size.

Note: This method assumes an 18-day dry period $(18/365 = 5\%)$

Table 7.8 Tank sizing based on 5% of rwh yield

5% of annual harvestable rainwater yield		
$Y_R = A \times h \times e \times \eta$		
I		
A	70	m²
e	0.9	
η	0.9	
h	557.49	mm
Y_R	31609.68	litres/yr
Y_R	31.61	m³/yr
5% Y_R	1.58	m³

Table 7.9 Tank sizing based on 5% of rwh demand

5% of annual non-potable water demand		
Daily Non-potable Water Demand		
$D_{p,d}$	60	(l/(p×d))
n (number of persons)	4	no.
$D_{N,d} = D_{p,d} \times n$	240	litres/day
$D_{N,a} = D_{N,d} \times 365$		
Annual Non-Potable Water Demand	87,600	litres/yr
	87.6	m3/yr
5% $D_{N,a}$	4.38	m³

The optimum rwh tank size according to this method is the lesser of the two values, 4.38m³, 1.58m³.

Nominal Rainwater Capacity:	1.58	m³

Worked Example 7.1 Summary

- The first conclusion is that rainwater harvesting is feasible at this site.
- The total annual rainfall depth is 557.49 mm giving an annual harvestable rainfall yield of 31.61 m³/annum.
- The total daily non-potable household water demand, based on a demand of 60 l/(p × d), 4 persons per household, is 240 litres/day (0.24m³/day).
- The total monthly and yearly household water demand is 7300 litres/month and 87,600 litres/annum respectively.
- Analysis of the monthly tabular method shows the distribution of rainfall across the year with the corresponding performance of the rainwater harvesting system.
- The annual supply coefficient (Harvestable rainfall yield/household water demand) is equal to 36%. This means that the rainwater harvesting system can supply 36% (31.61 m³/year) of the annual non-potable water demand for this household.
- The minimum optimum storage for this site is 1.58m³, or the closest commercially available tank available.

- The decision as to which is the optimum storage to provide at the site will ultimately depend on site constraints, availability of materials, cost etc.

Worked Example 7.2

For this example, we will take a similar residential house with 70m² contributing roof area located in Freetown, Sierra Leone, West Africa.

WORKED EXAMPLE 7.2 STAGE 1 HYDRAULIC FEASIBILITY STEP 1: ESTABLISH THE RAINFALL PROFILE FOR THE SITE LOCATION

What is the rainfall pattern for my site?

Table 7.10 shows average monthly rainfall (mm) figures taken from 10 years records for the site. This is in a tropical region with a distinct wet season dry season.

Table 7.10 Average monthly rainfall depths (mm) for Freetown, Sierra Leone

Jan	Feb	Mar	Apr	May	June	July	Aug	Sept	Oct	Nov	Dec	Total	Avg
15	14	16	68	198	509	1050	1104	780	330	131	48	4263	355.25

Plot the monthly rainfall and note any seasonal variations (Fig. 7.9).

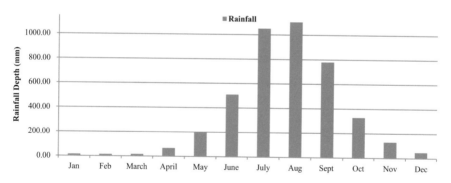

Fig. 7.9 Plot of monthly rainfall depths

This establishes a visual representation of the rainfall profile for the site in Freetown. For this particular site it can be seen that the rainfall profile is very different from the site in worked example 7.1. The total annual rainfall depth is 4263 mm. However, it can be seen from the rainfall pattern that there is a distinct dry season with very little rainfall from December – April. The average monthly rainfall for Freetown is 355.25 mm. There are 4 months when rainfall exceeded this average. Minimum rainfall occurs during February (14 mm) and maximum rainfall during August (1104 mm). The ratio of maximum to minimum rainfall is 78.85. This illustrates the variation in rainfall from dry season to wet season.

STEP 2: CALCULATE THE HARVESTABLE RAINWATER YIELD ($Y_{R,t}$)

What volume of rainfall can I potentially harvest from my site?

For the purpose of comparison, we will take a similar tiled roof on a residential house in Freetown Sierra Leone, with 70m² roof area draining to a single downpipe. There are typically three sources of water supply in Freetown, a communal water point, private water taps or private water vendors. Given the uncertain nature of the mains supply system the householder would like to investigate if it is feasible to utilise the rainwater harvesting system as their main source of household water year-round.

From Table 7.2 the surface yield coefficient (e) is 0.9. It is proposed to fit a first flush downpipe filter with a hydraulic treatment efficiency coefficient of 0.9.

$$Y_{R,t} = A \times h \times e \times \eta$$

$Y_{R,a}$ = 70 (m²) × 4263 (mm/yr) × 0.9 × 0.9 = 241,712.10 litres per annum (241.71 m³ per annum).

For this site we also have the monthly rainfall depths available. Therefore, we can repeat the above calculation using monthly rainfall depths to establish the harvestable rainwater yield per month.

We can now plot a profile of this analysis. This presents a useful graphical representation of any seasonal variation. The left vertical axis shows the monthly rainfall. The right vertical axis shows the harvestable rainwater yield. It can be observed that the harvestable rainfall yield is not consistent throughout the year (Fig. 7.10).

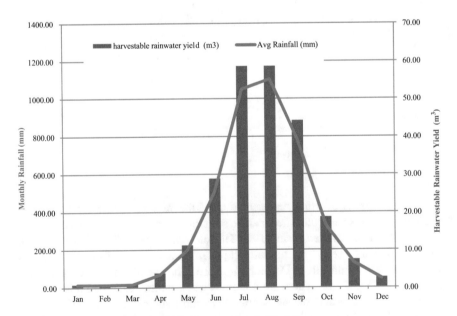

Fig. 7.10 Average monthly rainfall and harvestable rainfall yield

STEP 3: CALCULATE THE DEMAND PROFILE
What quantity of water is needed and when?

We can now establish the demand which we propose to supply from the harvested rainwater. In the previous example we investigated the feasibility of designing a rwh system to supply non-potable water only. With effective treatment the rwh system has the potential to supply potable water quality. In this second worked example we will adapt the methodology to design a system to replace the total water demand for a house. We will call this $D_{Total,t}$. We will adapt the methodology to suit. A previous study carried out in Freetown has established total household per capita consumption rate of 10 l/(p × d) for the location. There is an average house occupancy rate of 12 persons/house.

The total daily household water demand can be calculated;

$$D_{Total,d} = \frac{10l}{pxd} \times 12 \; persons = 120l/d \left(0.12 m^3/d\right)$$

The total annual household water demand ($D_{Total,a}$) expressed in litres per annum (l/a) can be calculated as follows:

$$D_{Total,a} = 120l/d \times 365 = 43,800l/a \left(43.8 m^3/a\right)$$

Similarly, we can calculate the monthly total water demand, $D_{Total,m}$ is equal to 3.65m³/month. It is assumed that this demand is constant throughout the year.

STEP 4: COMPARE AVAILABLE HARVESTABLE RAINWATER YIELD VERSUS DEMAND

We can summarise the design calculations in Table 7.11.

This now presents a prediction of how a rainwater harvesting facility may function at our site. We can see that demand exceeds supply for 4 months (December to March). The negative figure in column 5 shows the volume of additional water which is required to meet the demand during these months. Column 6 shows the supply coefficient, which is the percentage of demand which can be supplied from harvested rainwater. During January, February, and March the rwh system can only supply approximately 22 – 25% of the monthly demand. During April the harvested rainwater can supply all of the monthly water demand. For the months May to November inclusive, 100% of demand can be met from the rwh system. In particular during the months of July, August and September there is a surplus of over 1000%. In December the rwh system can supply 75% of monthly demand.

We can now establish the annual supply coefficient for this site as follows:

$$S(\%) = (241,712.1/43,780) \times 100 = 522\%$$

Examining the yearly supply figures, we can conclude that the potential harvestable rainfall can supply 522% of the total household water demand for this household.

Table 7.11 Tabular results for Freetown, Sierra Leone

Site location :

1	2	3	4	5	6
Month	Rainfall	Harvestable rainfall yield $Y_{R,m}$	Demand $D_{N,m}$	Deficit or Surplus $Y_{R,m}-D_{N,m}$	Supply coefficient $Y_{R,m}/D_{N,m} \times 100$
	(mm)	(m³)	(m³)	(m³)	%
Jan	15.00	0.85	3.65	−2.8	23%
Feb	14.00	0.79	3.65	−2.85	22%
March	16.00	0.91	3.65	−2.74	25%
April	68.00	3.86	3.65	0.21	106%
May	198.00	11.23	3.65	7.58	308%
June	509.00	28.86	3.65	25.21	791%
July	1050.00	59.54	3.65	55.89	1632%
Aug	1104.00	62.60	3.65	58.95	1716%
Sept	780.00	44.23	3.65	40.58	1212%
Oct	330.00	18.71	3.65	15.06	513%
Nov	131.00	7.43	3.65	3.78	204%
Dec	48.00	2.72	3.65	−0.93	75%
Total	**4263**	**241.71**	**43.78**	**197.94**	**552%**
Avg.	355.25				

Note: Column 5, a negative value indicates the volume of water required from an alternative water source to meet demand, a positive value indicated potential excess rainwater which could be stored

However, because the rainfall pattern is variable at this site, we also need to assess the monthly supply coefficients.

We can also plot the harvestable rainfall yield versus monthly demand (Fig. 7.11).

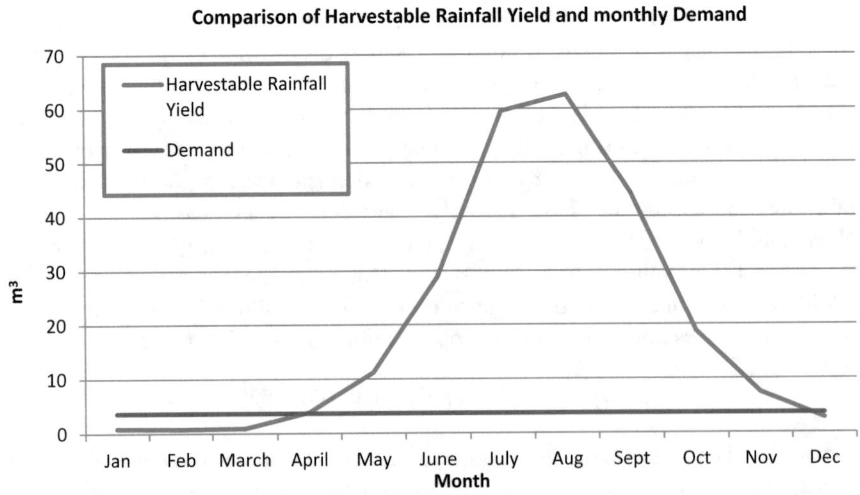

Fig. 7.11 Comparison of harvestable rainfall yield and monthly demand

This visual representation shows clearly the demand and supply profile. Demand is constant (indicated by the red line) throughout the year at 3.65 m³ per month. Supply (harvested rainfall yield) varies in relation to the corresponding rainfall profile. This graph illustrates visually the potential for storage of the excess rainwater during the wet season to meet excess demand during the dry season.

Finally, we can plot a comparison of harvestable rainfall yield with monthly demand and potential storage volume (Fig. 7.12).

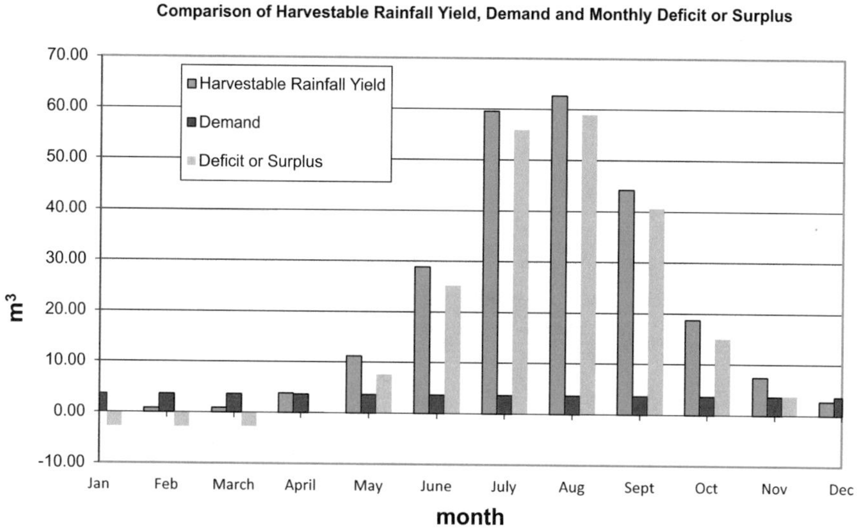

Fig. 7.12 Comparison of harvestable rainfall yield, rwh demand and rwh system deficit or surplus

This now gives us a more realistic prediction of how a rainwater harvesting system would function at this site location. This confirms the tabular assessment. We can clearly see the potential storage volume during the wet season which can supply the household water demand during the dry season. The rwh system from this analysis, with adequate storage can supply household water demand year-round. In fact, at this site, there is potential to store more water to meet additional demand in excess of the household demand at the site. This site could potentially provide a rainwater supply for a number of households. We will discuss this example in more detail in the next section.

Conclusion: Hydraulic efficiency justifies proceeding to Design Stage 2.
STAGE 2 STORAGE DESIGN

Finding the minimum size of tank using the tabular method
The performance characteristics of the site at Freetown, with variable rainfall, are suited to the intermediate method of storage design.

This method uses a standard tabular system. The results for this site are summarised in Table 7.12. We can also represent the results visually by plotting a graph

Table 7.12 Intermediate Tabular Storage Design Outputs

1	2	3	4	5	6	7	8	9
Month	Rainfall	Harvestable rainfall yield	Demand	Deficit or surplus	Supply coefficient	Cumulative harvestable rainfall yield	Cumulative Demand	Potential storage volume
		$Y_{R,m}$	$D_{N,m}$	$Y_{R,m}-D_{N,m}$	$Y_{R,m}/D_{N,m} \times 100$	$\sum Y_{R,m}$	$\sum D_{Total,m}$	
	(mm)	(m³)	(m³)	(m³)	%	(m³)	(m³)	(m³)
Jan	15.00	0.85	3.65	−2.80	23%	0.85	3.65	−2.80
Feb	14.00	0.79	3.65	−2.85	22%	1.64	7.30	−5.65
March	16.00	0.91	3.65	−2.74	25%	2.55	10.94	−8.39
April	68.00	3.86	3.65	0.21	106%	6.41	14.59	−8.18
May	198.00	11.23	3.65	7.58	308%	17.63	18.24	−0.61
June	509.00	28.86	3.65	25.21	791%	46.49	21.89	24.61
July	1050.00	59.54	3.65	55.89	1632%	106.03	25.54	80.49
Aug	1104.00	62.60	3.65	58.95	1716%	168.63	29.18	139.44
Sept	780.00	44.23	3.65	40.58	1212%	212.85	32.83	180.02
Oct	330.00	18.71	3.65	15.06	513%	231.56	36.48	195.08
Nov	131.00	7.43	3.65	3.78	204%	238.99	40.13	198.86
Dec	48.00	2.72	3.65	−0.93	75%	241.71	43.78	197.94
Total	4263	241.71	43.78	197.94	552%			
Avg.	355.25							

of cumulative demand versus cumulative harvestable rainfall supply. This clearly shows the supply deficit during the months January to April inclusive and December. The supply line crosses the demand line between May and June. For the remainder of the year the supply line is above the demand line confirming our previous analysis.

Table 7.12 allows us to quantify the storage requirements by using a cumulative analysis. Column 7 shows the cumulative monthly harvestable rainfall yield. Column 8 shows the cumulative rwh demand. Column 9 shows the potential storage volume of rwh which is equal to column 7 minus column 8. Column 9 now allows us to interrogate the predicted performance of the rwh system. An analysis of column 9 shows two items. The largest negative number gives us the minimum storage requirement to meet 100% of household water demand. For this example, the minimum rwh storage volume to meet demand during the first 4 months is 8.39m³. The maximum positive number gives us the largest possible volume of rainwater which could potentially be stored from this system. For this example, it is equal to 198.86m³. We can plot these results in Figure 7.13.

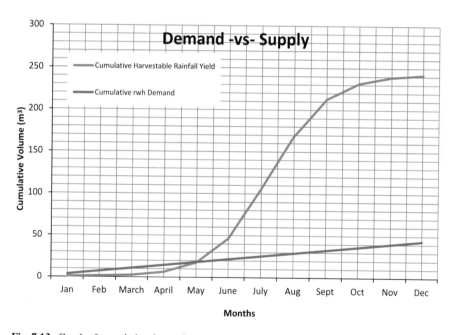

Fig. 7.13 Graph of cumulative demand versus supply

Summary

The total annual rainfall depth is 4263 mm giving an annual harvestable rainfall yield of 241.71 m³/annum. The total daily household water demand, based on a household characteristics of 10 l/(p × d), 12 persons per household, is 120 litres/day (0.12m³/day). The total monthly and yearly household water demand is 3648 litres/

month and 43,800 litres/annum respectively. The annual supply coefficient (Harvestable rainfall yield/household water demand) is equal to 552%. Analysis of monthly tabular method shows the variability of rainfall with a distinct wet season and dry season. A minimum storage volume of 8.39m^3 is required to meet household water demand during the dry season. During the wet season the rainwater harvesting system can meet 100% of household water demand. A maximum volume of 198.86 m^3 of rainwater could potentially be stored at the site. This storage volume could potentially supply the annual household water demand of an additional 4.5 households (198.86/43.8). The decision as to which is the optimum storage to provide at the site will ultimately depend on site constraints, availability of materials, cost etc. Carrying out a hydraulic feasibility study as discussed in this chapter will allow the user to make an informed decision on investing in a rainwater harvesting system.

Chapter 8
The Economics of Rainwater Harvesting

NO ONE VALUES RAINWATER!

8.1 Introduction

How much water does a typical household use per day?
The International Water Association (IWA) is a worldwide network of professionals which aims to exchange scientific and professional information covering many

© Springer Nature Switzerland AG 2021
L. McCarton et al., *The Worth of Water*,
https://doi.org/10.1007/978-3-030-50605-6_8

aspects of the water cycle (www.iwa-network.org). They provide a useful resource website which collates information covering abstraction, consumption, tariff structure and regulation of water services globally from 39 countries and 198 cities worldwide (www.waterstatistics.org). Household consumption per person varies widely across regions. For example it is recorded as 14 litres per person per day in Kampala, Uganda to 330 litres per person per day in Buenos Aires, Argentina. In the economic analysis local data should be accessed, if available.

What is the cost of water for a typical household?

The International Benchmarking Network for Water and Sanitation Utilities (IBNET) provide direct access to the world's largest database of water and sanitation tariffs performance data (www.tariffs.ib-net.org). The IBNET tariff database is a joint product of Global Water Intelligence (GWI) and the International Benchmarking Network of the World Bank (IBNET). The database provides data on water and wastewater utility charges across 206 countries. It is based on data from the annual Global Water Intelligence survey which collates and publishes global tariffs for water and wastewater. A review of the most recent data shows the global average for a combined water and wastewater tariff was $2.04/m³. The introduction of water charges in Turkmenistan leaves the Republic of Ireland as the only country in the global survey not to apply a domestic charge for water. Table 8.1 reproduces the global combined water and wastewater tariffs. The global wastewater tariff increased faster than the average water tariff.

Figure 8.1 shows a breakdown of the components of the global average combined water tariff.

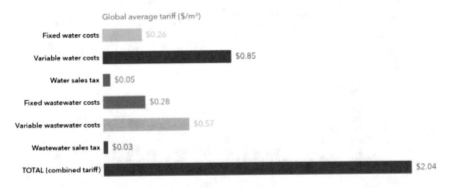

Fig. 8.1 Breakdown of combined water and wastewater tariff. (Source: www.globalwatersecurity.org)

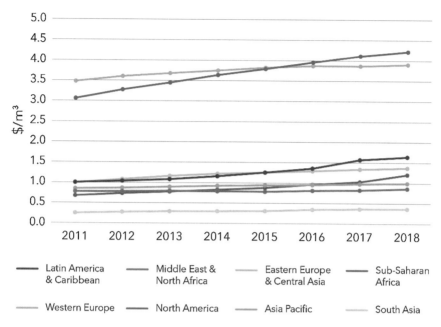

Fig. 8.2 Global variation in average combined water and wastewater tariff. (Source: www.global-watersecurity.org)

Figure 8.2 shows the variation in the average water tariff by region, 2011–2018. In Western Europe and North America, the need to upgrade aging infrastructure is a primary driver of increased tariffs. In other regions the influence of climate change is a major factor. In 2018 Cape Town (South Africa), due to a water shortage caused by prolonged drought, was forced to implement a 390% increase in combined water & wastewater tariff while limiting the maximum daily per capita allowance to 50 litres.

Table 8.1 Combined Water and Wastewater Tariffs, based on the charge for a household of 4 persons using 15m³/month (adapted from GWI Global Survey, 2019)

Country	($/m³)	Country	($/m³)	Country	($/m³)	Country	($/m³)	Country	($/m³)	Country	($/m³)
Bermuda (UK)	6.06	Liechtenstein	3.12	Dominica	1.78	Zambia	0.87	Argentina	0.56	Nepal	0.24
Curaçao	5.88	Poland	3.06	UAE	1.77	Morocco	0.87	Mozambique	0.53	Laos	0.24
Cayman Islands (UK)	5.86	Cyprus	3.02	Somaliland	1.73	Saint Kitts and Nevis	0.86	Timor-Leste	0.52	Burundi	0.23
Nauru	5.83	Sweden	3.01	Uganda	1.65	Russia	0.85	Bolivia	0.49	Guinea-Bissau	0.23
Kiribati	5.72	Greenland	2.90	Seychelles	1.64	Djibouti	0.83	China	0.49	Kazakhstan	0.21
Switzerland	5.66	Belize	2.87	Romania	1.63	Burkina Faso	0.82	Côte d'Ivoire	0.49	Bangladesh	0.21
Monaco	5.61	Croatia	2.52	Colombia	1.59	Mexico	0.82	Central African Rep.	0.47	Lebanon	0.21
Malta	5.56	Saint Lucia	2.48	Lithuania	1.56	Vanuatu	0.81	Eritrea	0.46	Mauritius	0.21
Australia	5.39	South Africa	2.47	St Vincent/Gr'dines	1.54	Turkey	0.79	Tunisia	0.43	Dominican Republic	0.21
Luxembourg	5.24	Marshall Islands	2.39	Qatar	1.51	Ukraine	0.78	Mali	0.41	Gambia	0.20
Belgium	5.24	Spain	2.36	Bulgaria	1.49	Benin	0.78	Afghanistan	0.40	Chad	0.18
Cape Verde	5.17	Estonia	2.34	Costa Rica	1.48	Rwanda	0.78	Trinidad and Tobago	0.39	Sri Lanka	0.18
Netherlands	4.92	Brazil	2.21	Greece	1.48	Serbia	0.76	Niger	0.38	Turkmenistan	0.17
Bahamas	4.91	Latvia	2.17	Lesotho	1.44	Paraguay	0.74	Zimbabwe	0.38	Malaysia	0.17
Norway	4.81	Italy	2.17	Liberia	1.42	Montenegro	0.73	Armenia	0.38	Fiji	0.17
eSwatini	4.79	Singapore	2.15	Chile	1.32	Moldova	0.71	Taiwan	0.37	Cambodia	0.17
Anguilla (UK)	4.54	Samoa	2.12	Botswana	1.30	Malawi	0.70	Iraq	0.37	Iran	0.14
Austria	4.47	Israel	2.08	Grenada	1.21	Cameroon	0.69	Thailand	0.35	Saudi Arabia	0.13

(continued)

Country	Value	Country	Value	Country	Value	Country	Value	Country	Value	Country	Value
Finland	4.42	Antigua/Barbuda	2.08	Bahrain	1.21	Senegal	0.68	Myanmar	0.34	Ethiopia	0.13
United States	4.40	Portugal	2.04	Solomon Islands	1.16	Belarus'	0.68	Republic of Congo	0.34	India	0.13
United Kingdom	4.38	Slovenia	2.04	Kenya	1.12	Mongolia	0.67	El Salvador	0.32	Georgia	0.12
Maldives	4.33	Japan	2.03	Ghana	1.06	Papua New Guinea	0.66	Nicaragua	0.31	Honduras	0.11
France	3.92	Jamaica	2.03	Panama	1.06	Nigeria	0.65	Philippines	0.30	Egypt	0.11
British Virgin Islands	3.86	Palestine	2.01	Gabon	1.03	Ecuador	0.64	Vietnam	0.30	Kyrgyz Republic	0.11
Czech Republic	3.67	Hungary	1.98	South Korea	1.00	Guatemala	0.64	Azerbaijan	0.29	Uzbekistan	0.08
Aruba	3.67	Equatorial Guinea	1.95	Albania	0.99	Togo	0.61	Angola	0.28	Brunei	0.08
Canada	3.44	Montserrat	1.93	Micronesia	0.99	Jordan	0.60	Indonesia	0.28	Pakistan	0.07
New Zealand	3.29	Oman	1.89	Syria	0.98	North Macedonia	0.59	Suriname	0.28	Tajikistan	0.07
Slovakia	3.15	Barbados	1.85	Peru	0.96	Guyana	0.58	Algeria	0.25	Cuba	0.02
Namibia	3.14	Uruguay	1.82	Hong Kong SAR	0.91	Kuwait	0.58	Madagascar	0.24	Ireland	0.00

8.2   Economic Analysis Tools

Economic Analysis Tools

There are a number of recognised financial tools which can be used to evaluate the economic performance of a potential investment.

- **Simple Payback**
- **Net Present Value**
- **Equivalent Annual Cost**

Simple Payback

The simple payback period is used in financial and capital budgeting to determine the length of time for an investment to reach a breakeven point. This is calculated by dividing the initial cost of the investment by the annual savings generated.

$$\text{PAYBACK PERIOD} = \frac{C_0}{\text{NAS}} \tag{8.1}$$

Where:

C_0 = the initial investment cost
NAS = net annual savings

Example: A householder is considering investing in a solar panel system with an initial capital outlay of €5000. The annual power output from the solar system will result in a reduction in monthly electricity charges of €100 per month. For this example the payback period is;

$$\text{PAYBACK PERIOD} = \frac{5{,}000}{100 \times 12}$$

For this example, the simple payback would be 4.2 years, i.e. It will take 4.2 years for the annual savings to equal the initial investment costs.

Advantages:
- Very simple method

Disadvantages:
- The method takes no account of the time value of money and neither does it take account of the earnings after the initial investment is recouped.

Life-cycle costing

A number of more detailed accounting techniques have been developed which evaluate the cost of ownership of an asset over its entire lifetime. This includes all life time costs including design, construction, operation, maintenance, decommissioning and disposal costs. Within these methods the relationship between the value of money and time is evaluated.

Discount Rate (r %): A private or public investor typically borrows the money to pay for projects with an expectation that the long term benefit of the scheme will pay for the repayment of the loan. Any financial benefits accruing after the initial loan has been repaid will constitute a return (profit) on the investment. This introduces a concept known as "time value of money". This is the long term value of money which is linked to the interest rates associated with the loan. This is termed the "discount rate". The setting of this rate is a complex process and is beyond the scope of this text. For the purposes of the case studies in this chapter a standard discount rate of 4% is assumed for all calculations.

Present Value (PV): The concept of present value describes how much a future sum of money is worth today.

$$PV = \frac{F}{(1+r)^n}$$

(8.2)

Where

F = future value
r = discount rate (also known as periodic rate of return)
n = number of periods

Example: You want to invest money in an interest account today to provide a cash sum of €10,000 in 10 years time on your retirement. You know you can get an interest rate of 5%. How much should you invest today.

The present value formula tells us that: $PV = €10,000/(1 + 0.05)^{10} = €6139.13$

This means that €6139.13 will be worth €10,000 in 10 years if you can earn 5% each year. In other words, the present value (PV) of €10,000 in this scenario is €6139.13.

The present value concept is one of the most fundamental in the world of finance. It accounts for the "time value of money".

Net Present Value (NPV)

Consider the funding of a project. If a project purchases an asset that has a lifetime of T years, a net benefit (*NB*) can be drawn from the project each year of its operation. At the start of the project the sum representing present value *(PV)* is borrowed at r % to fund the entire project.

Each year a cash flow (CF) is generated which is equivalent to the cost of ownership less the expected savings each year. A table can be drawn up as follows (Table 8.2):

Table 8.2 Net Present Value

Time (t) years	0	1	2	3	n
Cash Flow (CF)	CF_0	CF_1	CF_2	CF_3	CF_n

The Net Present Value can be calculated as follows:

$$NPV(r,n) = \sum_{t=0}^{n} \frac{CF_t}{(1+r)^t} \qquad (8.3)$$

Where CF_t represents the net cash flow in year t, r is the discount rate and n, represents the lifetime of the project. The resulting sum of discounted cash flows equals the projects net present value. For the above table it would be calculated as follows:

$$NPV = CF_0 + \frac{CF_1}{(1+r)^n} + \frac{CF_2}{(1+r)^2} + \frac{CF_3}{(1+r)^3} + \ldots \frac{CF_n}{(1+r)^n} \qquad (8.4)$$

The cash flows in each year may be positive or negative. For example, the initial cash flow, CF_0 is a negative number representing the initial outlay necessary to get the project started. CF_1 represents the net cash flow in year 1, which will equate to the income and/or savings generated less any lifetime operating and maintenance costs.

The basic accounting rule says that an investment is recommended when the NPV > 0. This will occur when the following occurs:

$$CF_0 < \frac{CF_1}{(1+r)^n} + \frac{CF_2}{(1+r)^2} + \frac{CF_3}{(1+r)^3} + \ldots \frac{CF_n}{(1+r)^n} \qquad (8.5)$$

Simply stated the NPV rules are:

NPV >€0 Invest

NPV <€0, Do Not Invest

The NPV in accountancy terms equates to the amount of additional value created by an investment.

Equivalent Annual Cost
In this technique the cost of ownership over the lifetime of an asset is annualised and the equivalent annual cost (EAC) is calculated. This is the annual cost of owning, operating, maintaining and disposing of an asset over its lifetime. It is determined by dividing the net present value of the asset purchase, operations, maintenance and disposal cost divided by the present value annuity factor.

$$\text{Present Value Annuity Factor} = \frac{1-(1+r)^{-n}}{r} \qquad (8.6)$$

Where:

r = discount rate
n = number of periods

USING THE ECONOMIC TOOLS TO EVALUATE RAINWATER HARVESTING INSTALLATIONS

We can now apply some of the economic tools discussed in the previous pages to determine the economic performance of a number of rainwater harvesting (rwh) applications. The approach for all case studies will be to use a comparison of a user with mains water supply only compared to a user with mains water and rwh system installed. Figure 8.3 and Fig. 8.4 illustrate the information required to carry out the analysis.

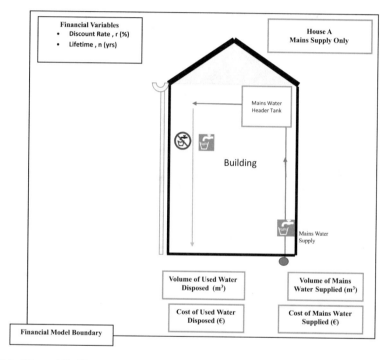

Fig. 8.3 "House A" – House supplied by mains water only

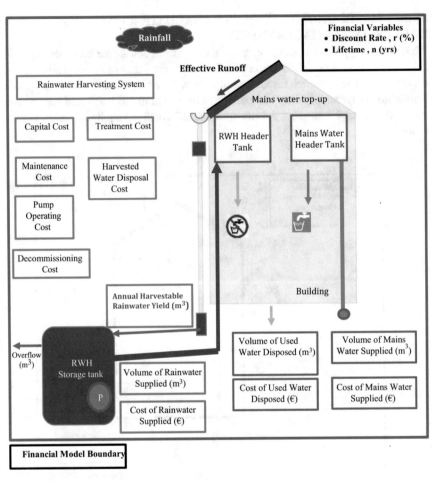

Potable (Drinking Water) in accordance with ISO 7010:E015

Non - Potable Water in accordance with ISO 7010:P005

Fig. 8.4 "House B" – House supplied by rwh system and mains water

8.3 A Methodology for the Economic Appraisal of rwh Systems

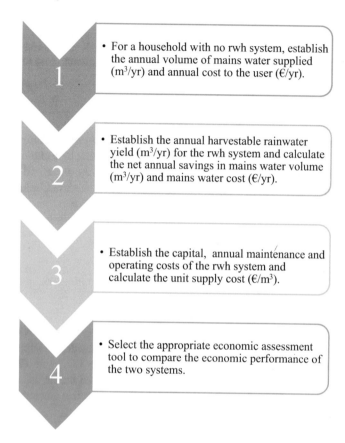

Fig. 8.5 Outlines the methodology for the economic appraisal of rwh systems

Worked Example 8.1

In Chap. 7 we carried out a hydraulic design of a rainwater harvesting system for a typical house. We can now use these design figures to compare the economics of the following:

- **House A – <u>mains water</u> supply only.**
- **House B – <u>rwh system</u> plus <u>mains water</u> supply**

From Chap. 7, the design characteristics for worked example 7.1 were as follows:

- Four person household, in a location with an annual average rainfall depth of 557.49 mm, a tiled roof rwh catchment area of 70 m².

- Assume a non-potable water demand per person of 60 l/(p × d) which comprises of 50 l/(p × d) for WC flushing and 10 l/(p × d) for miscellaneous external garden use.

1 • **For a household with no rwh system, establish the annual volume of mains water supplied (m³/yr) and annual cost to the user (€/yr).**

For this example we will assume a per capita household water consumption rate of 150 l/(p × d), with 4 persons per household. This gives a total household water consumption of 219,000 litres/yr (219m³/yr). Table 8.3 shows the water and wastewater tariffs used in this example. Using these the annual cost to the household is €308.79 for water (based on €1.41/m³).

2 • **Establish the annual harvestable rainwater yield (m³/yr) for the rwh system and calculate the net annual savings in mains water volume (m³/yr) and mains water cost (€/yr).**

From Chap. 7 we have already established the annual harvestable rainwater yield as 31.61 m³/yr. The annual household mains water volume needed to supply the estimated household water consumption of 219m³/yr. will be 187.39 m³ (219–31.61). The corresponding mains water cost for this volume, based on the assumed tariffs will be €264.22/yr. Therefore, the net annual savings for this household will be €44.57/yr. Table 8.3 summarises this analysis.

Table 8.3 Mains water supplied and costs to supply

	Water	Used water (wastewater)	Combined
Water & used water tarriffs (€/m³)	1.41	2.04	3.45
House A – mains water system only			
Annual Mains Supply 219 m³	€ 308.79	€ 446.76	€ 755.55
House B – mains water system and rwh system			
Annual Mains Supply 187.39 m³	€ 264.22	€ 446.76	€ 710.98

Since the wastewater volumes discharged from the two household will be equal, we will only consider the water supply costs in our economic comparison.

3 • **Estimate the capital, maintenance and operating costs of the rwh system (€).**

Capital Costs

The capital costs of any rwh system typically comprise two main components as follows:

- rwh System
- rwh Storage Tank

(i) **rwh System**

Most domestic systems comprise a number of filters (downpipe and underground), a submersible pump with float switches & floating suction filter, rainwater header tank with internal fittings, a mains water top up device and air gap and a rwh electronic control unit. Table 8.4 presents the typical costs for a rwh domestic installation in Ireland.

Table 8.4   Typical rwh system component costs for a domestic installation in Ireland

RWH Component	Cost (€)
Filters (downpipe, underground)	700
Rainwater Storage tank fittings (inlet, overflow)	145
Submersible pump with float switches & floating suction filter	875
Rainwater header tank with fittings	325
Mainswater top up with piggy back plug and air gap	155
rwh control unit	250
Total (ex VAT)	**€2450**

These costs are independent of storage volumes. Typical labour costs for new build domestic systems are of the order of €1000 (based on Irish construction rates 2019).

(ii) **rwh Storage Tank**

Costs are variable depending on the tank size and material and whether it is above or below ground. HDPE tanks costs vary from €650/m³ storage for underground to €300/m³ for above ground tanks. Typical costs for concrete tanks vary from €1000–€2500.

Total rwh Capital Costs

The total capital cost for a new build domestic rwh system will typically be in the range €4450–€5950. These costs are consistent with a study carried out in 2007 by Roebuck in the UK which found that the capital costs of new build rwh systems ranged from 2500–5500£Stg.

Operating Costs

The main operating costs of a rwh system consist of electricity charges. Based on average pump use age of 0.5 hrs per day, the annual electricity charge based on 17 cents per kilowatt hour is €23.14.

Maintenance Costs

Table 8.5 shows a typical maintenance schedule for a domestic rwh system. Annualising the maintenance costs presents an annual maintenance cost of €124.53.

Table 8.5 shows a typical maintenance schedule for a domestic rwh system

Maintenance Item	Frequency	Estimated Cost (€)	Annualised Cost (€)
Replace mains top up solenoid valve	7.5 years	142	18.93
Replace submersible pump	10 years	600	60.00
Replace mains top-up level switch	12.5 years	70	5.60
Replace coarse filter	15 years	350	23.33
Replace electronic controls	15 years	250	16.67
Subtotal			**€124.53**

For the financial model a discount rate of 4% will be selected. A lifetime of 20 years will be selected. Figure 8.6 presents a schematic of the mains water system annual performance costs to the household for House A. The combined annual cost of mains water supply and used water disposal, based on the tariff assumptions presented in Table 8.3, is €755.55/yr. These are the annual costs without any rwh system installed. Figure 8.7 presents a similar schematic of the mains water supply, used water disposal and rwh system annual costs to the householder for House B.

> **4**
> • **Select the appropriate tool to compare the economic performance of the two systems.**

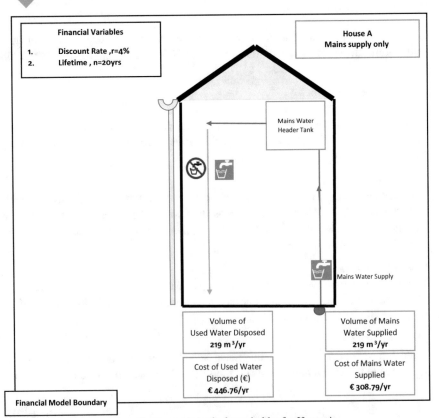

Fig. 8.6 Annual water and used water costs to the householder for House A

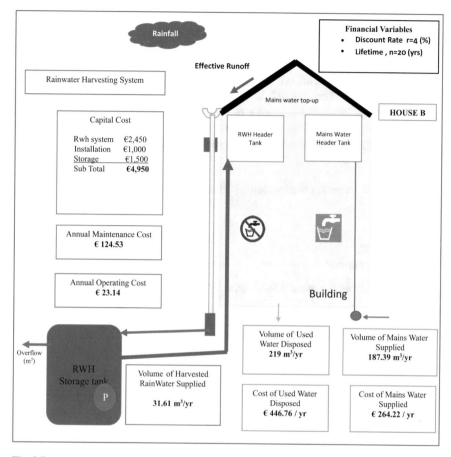

Fig. 8.7 Annual water and used water costs to the householder for House B

(i) Simple Payback

The simple payback period is used in financial and capital budgeting to determine the length of time an investment needs to reach a breakeven point.

For House B the payback calculation is as follows:

$$Capital \;\; Cost = \;\; 4,950$$

Net Annual Savings = Annual saving on mains water costs less (annual operating plus maintenance costs) = 44.57 − (124.53 + 23.14) = €−103.67

It becomes immediately apparent that there is no payback since the net annual maintenance and operating costs are greater than the annual value of mains water saved. This simple analysis does not however tell us anything about how House A compares with House B over the lifetime costs of the system since it does not take into account the time value of money. It also does not allow us to compare the unit costs of the rwh system with the unit costs for the mains water system. To do this we need to apply some of the more detailed financial tools.

(ii) Equivalent Annual Cost

In this technique the cost of ownership over the lifetime of an asset is annualised and the equivalent annual cost (EAC) is calculated. This is the annual cost of owning, operating, maintaining and disposing of an asset over its lifetime. It is determined by dividing the net present value of the asset purchase, operations, maintenance and disposal cost divided by the present value annuity factor.

$$\text{Present Value Annuity Factor} = \frac{1-(1+r)^{-n}}{r} \qquad (8.7)$$

Where:

r = discount rate
n = number of periods

For this analysis the present value annuity factor can be calculated as;

$$\text{Present Value Annuity Factor} = \frac{1-(1+0.04)^{-20}}{0.04} = 13.59$$

Table 8.6 shows a blank template which we can utilise for this analysis.

Table 8.6 Template to calculate annual equivalent cost of the rwh system

Finance Variables	Discount factor, r (%)	
	No. of years, n	
	PV annuity factor $\dfrac{1-(1+r)^{-n}}{r}$	
Annual Equivalent Costs	Capital Costs	
	Annual Equivalent of Capital Costs	
	Total Annual Equivalent Cost (including maintenance & operating costs)	
	Total Present Value of rwh installation	
RWH production	Annual Harvestable Rainfall Yield, Y_R,a (m³/yr)	
	Percentage Non-potable Demand met (%)	
	RWH total yield over lifetime (m³)	
Volumetric Unit Costs	RWH total unit production cost (€/m³)	
	RWH operation & maintenance unit production cost (€/m³)	
	Mains Water Volumetric Tariff (€/m³)	
Annual Mains Water Costs	Annual Household Mains Water Cost (€)	
	Annual Household Mains Water Cost with rwh system, (€)	
	Annual Net Savings on Water Charges (€)	
Cost Benefit Ratio	Present Value of Benefits	
	Cost Benefit Ratio	

Table 8.7 presents a summary of the results for worked example 8.1.

Table 8.7 Results of annual equivalent cost analysis for worked example 8.1

Financial Variables	Discount factor, r (%)	4%
	No. of years, n	20 years
	PV annuity factor $\dfrac{1-(1+r)^{-n}}{r}$	13.59
Annual Equivalent Costs	Capital Costs	€ 4950
	Annual Equivalent Capital Costs	€ 364
	Total Annual Equivalent Cost (including maintenance & operating costs)	€512
	Total Present Value of rwh installation	€6957
RWH production	Annual Harvestable Rainfall Yield, $Y_{R,a}$ (m³/yr)	31.61 m³/ yr
	Percentage Non-potable Demand met (%)	36%
	RWH total yield over lifetime (m³)	632 m³
Volumetric Unit Costs	**RWH total unit production cost (€/m³)**	**16.20**
	RWH operation & maintenance unit production cost (€/m³)	**4.67**
	Mains Water Volumetric Tariff (€/m³)	**1.41**
Annual Mains Water Costs	Annual Household Mains Water Cost – House A (€)	308.79
	Annual Household Mains Water Cost – House B, (€)	264.22
	Annual Net Savings on Water Charges (€)	44.57

Financial Variables

We have selected a discount rate of 4%, and a 20 year lifetime. The Present Value Annuity Factor is 13.59.

Annual Equivalent Costs

The capital cost of the rwh system is €4950. The annual equivalent cost of servicing this capital at 4% per annum over 20 years can be calculated by dividing the capital costs by the Present Value Annuity Factor = 4950/13.59 = €364. To service the initial rwh capital investment would require €364 per annum. To calculate the total annual equivalent costs we need to add the annual operating and maintenance cost. The annual operating and maintenance cost = 124.53 + 23.14 = €147.67. Total annual equivalent costs are calculated as 364 + 147.67 = €512. The present value of the rwh system = 512 × 13.59 = €6957.

RWH Production

The annual harvestable rainwater yield is 31.61 m³/yr. Over 20 years the rwh yield will be 31.61 × 20 = 632 m³.

Volumetric Unit Costs

The rwh unit costs can be calculated by dividing the total annual equivalent cost by the annual rainwater yield = 512/31.61 = €16.20/m³. **Therefore, for this installation, over a 20 year lifespan, it will cost €16.20 to supply each 1m³ of harvested rainwater. The corresponding operational unit cost** can be calculated by dividing the annual operating and maintenance cost by the annual yield as follows; 124.53

+ 23.14 = 147.67/31.61 = **€4.67/m³**. The corresponding mains unit costs (water tariff) for this example is €1.41/m³.

Annual Mains Water Cost

For House A the household mains water cost is €308.79. For House B the mains water cost will be €264.22. **The net annual cost saving to House B on mains water charges is €44.57 per year.**

One of the advantages of the annual equivalent cost method is that it allows us to quickly compare the unit cost of rwh system with the mains water tariffs. This methodology can then be used to evaluate the impact of any measures such as providing capital grants as an incentive to install rwh systems. For example if a 50% capital grant was provided to House B, this would reduce the rwh unit costs to €10.43/m³.

(iii) **Net Present Value (NPV)**

The Net Present Value of both mains water only and mains water plus rwh system can be calculated from Eq. 8.3.

Table 8.8 shows the NPV calculations. A detailed break down of the calculations for House A and House B is as follows;

Row 1 shows years 0 to 20

Row 2 shows the estimated yearly cash flows for House B - a house fitted with a rwh system and a mains water supply system.

> Year 0 Cash Flow = −€4950 (minus indicates capital investment in year 0 of €4950 for the rwh system)
> Year 1 – Year 20 net cash flow = mains water cost for rwh house of €264 plus annual operating & maintenance cost for the rwh system of €147.68 = −€412 (minus shows an outgoing net cash flow).

Row 3 shows the corresponding estimated cash flows for House A - a house fitted with a mains water supply only and no rwh system. In this case there is no capital investment cost and an annual mains water cost of €309.

> Year 1 – Year 20 cash flow is equal to the mains water costs of €309. Again minus indicates an outgoing net cash flow.

Row 4 shows the discount rate selected for this example of 4%.

Row 5 shows present value calculations for House B

> Year 0 = −€4950 (equivalent to the capital cost of the rwh system)

$$\text{Year } 1 = \frac{CF_1}{(1+r)^n} = \frac{-412}{(1+0.04)^1} = -\text{€}396$$

$$\text{Year } 2 = \frac{CF_1}{(1+r)^n} = \frac{-412}{(1+0.04)^2} = -\text{€}381$$

Row 6 shows present value calculations for House A

Year 0 = 0

$$\text{Year } 1 = \frac{CF_1}{(1+r)^n} = \frac{-309}{(1+0.04)^1} = -297$$

$$\text{Year } 2 = \frac{CF_1}{(1+r)^n} = \frac{-309}{(1+0.04)^2} = -285$$

Row 7 shows the NPV calculation for House B as follows:

$$NPV(r,n) = \sum_{t=0}^{n} \frac{CF_t}{(1+r)^t} = -4950 + \frac{-412}{(1+0.04)^1} + \frac{-412}{(1+0.04)^2}$$
$$+ \frac{-412}{(1+0.04)^3} + \dots \frac{-412}{(1+0.04)^{20}}$$

NPV House B = − €10,548

Row 8 shows a similar NPV calculation for house A as follows:

$$NPV(r,n) = \sum_{t=0}^{n} \frac{CF_t}{(1+r)^t} = 0 + \frac{-309}{(1+0.04)^1} + \frac{-309}{(1+0.04)^2}$$
$$+ \frac{-309}{(1+0.04)^3} + \dots \frac{-309}{(1+0.04)^{20}}$$

NPV House A = (−) €4917

The Difference over 20 years is equal to (−) €6352.

Discussion

We now have a more complete picture of the predicted economic performance of the rwh system for this case study. The simple payback method shows effectively that there is no payback to the consumer since the net annual maintenance and operating costs are greater than the annual value of mains water saved. The second economic tool is the annual equivalent cost method. The results from this analysis (Table 8.7) show that the unit rwh supply costs are of the order of €16.20/m³. This is far in excess of the mains water costs to the consumer. Finally, the NPV method which takes into account the time value of money (Table 8.8) shows that over the lifetime of the rwh system it will cost €6352 more than mains water supply. Therefore, for this householder there is no financial incentive to install a rwh system without any subsidy. Figure 8.8 further illustrates the comparison between the results for House A and House B. This is typical of many European countries where subsidies are applied to encourage householders to install rwh system. For example, in German the authorities apply a "rain tax". This is a charge based on the catchment area of your roof. Householders who have a rwh system installed get a rebate on this

Table 8.8 Net Present Value (NPV) Analysis comparing House A with House B

Inputs

		Year	0	1	2	3	4	5	6	7	8	9	10	11	12	13	14	15	16	17	18	19	20
Row 1																							
Row 2	Cash flows for House B - (rwh system and mains water supply)		-4950	-412	-412	-412	-412	-412	-412	-412	-412	-412	-412	-412	-412	-412	-412	-412	-412	-412	-412	-412	-412
Row 3	Cash flows for House A - (mains water only)		0	-309	-309	-309	-309	-309	-309	-309	-309	-309	-309	-309	-309	-309	-309	-309	-309	-309	-309	-309	-309
Row 4	Discount rate	4.0%																					

Present Value Calculations

		Year	0	1	2	3	4	5	6	7	8	9	10	11	12	13	14	15	16	17	18	19	20
Row 5	PV Cash flows for House B - (rwh system and mains water)		-4950	-396	-381	-366	-352	-339	-326	-313	-301	-289	-278	-268	-257	-247	-238	-229	-220	-211	-203	-196	-188
Row 6	PV Cash flows for House A - (mains water only)		0	-297	-285	-275	-264	-254	-244	-235	-226	-217	-209	-201	-193	-185	-178	-171	-165	-159	-152	-147	-141

Discounted Cash Flow

		Year	0	1	2	3	4	5	6	7	8	9	10	11	12	13	14	15	16	17	18	19	20
Row 7	Cumulative Discounted N PV Cash flows for House B- (rwh system and mains water)		-4950	-5347	-5727	-6094	-6446	-6784	-7110	-7423	-7724	-8013	-8291	-8559	-8816	-9064	-9301	-9530	-9750	-9961	-10,165	-10,360	-10,548
Row 8	Cumulative Discounted N PV Cash flows for House A - (mains water only)		0	-297	-582	-857	-1121	-1375	-1619	-1853	-2079	-2296	-2505	-2705	-2898	-3083	-3262	-3433	-3598	-3757	-3909	-4056	-4197

tax. This analysis has only focused on the costs to the householder. The analysis could be repeated to include the cost savings to the municipal water supply company to give a full picture of the economics of rwh.

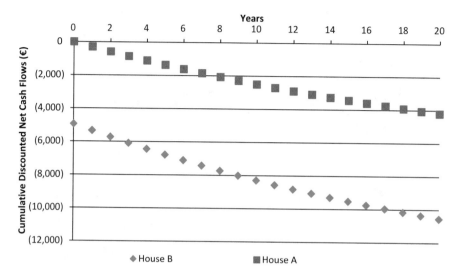

Fig. 8.8 Cumulative discounted net cash flows for House A and B

Worked Example 8.2

In developing countries around the world there is often a lack of access to a reliable water source. In this context the practice of rainwater harvesting can provide a safe source to meet all household water demands. In Chap. 7, (worked example 7.2) we examined the hydraulic performance of a household rwh system in Freetown, Sierra Leone, West Africa. We can now use this example to assess the economic performance of this type of system (Fig. 8.9).

To recap, the design characteristics are as follows:

- Annual harvestable rainfall yield 241.71 m³/year.
- Annual household water demand 43.8 m³/year, based on a daily demand of 10 l/ (p × d) × 12 persons = 120 litres/day.
- A minimum storage volume of 10m³ will meet 100% of household water demands for the year.

We can apply a similar methodology to worked example 8.1 to evaluate the economic performance of this rwh system.

What is the cost of mains water?

Access to water in Freetown is via one of three possible supply options. The first option is to collect water daily at a communal borehole fitted with a handpump. The second option is via a private water tap supplied from the mains water system. The third option is to purchase water from private water vendors. Table 8.9 shows the water consumption tariff for each supply option. The corresponding estimated annual household water costs for each supply option are also listed. We will use the

Fig. 8.9 Typical low cost rwh system in Sierra Leone

same comparison as for the previous example, where House A has no rwh system fitted, and House B has a rwh system. For this example House B can meet its annual water needs from the rwh system, therefore public water supply costs are negligible.

Table 8.9 Consumption Tariffs and estimated rwh system costs for Freetown, Sierra Leone

Water consumption tariffs Freetown, Sierra Leone			
Consumption Tariff (€/m³)	Communal Borehole	Private Water Tap	Water Vendor
	0.70	1.40	6.50
House A – public water supply only			
Public water demand 43.8 m³/yr	€ 31	€ 61	€ 285
House B – public water supply and rwh system			
Public water demand 0 m³/yr	–	–	–

What are the costs of a Domestic Rainwater Harvesting System?
Capital Costs

The rwh system for Freetown will not involve high cost proprietary components. The capital costs will comprise of guttering, first flush device, fittings and pipework and storage tank. The storage tank will comprise the largest percentage of the capital costs. Table 8.10 summarises the estimated capital costs in euros for three standard rwh storage tank options. Prices are from the authors work in Sierra Leone (2018). Ferrocement and blockwork tank options have cheaper material costs but higher labour costs. Plastic tanks are readily available and require lower labour costs to fit and maintain.

Table 8.10 Typical rwh storage tank costs for Freetown, Sierra Leone

10m³ rwh storage tank	Ferrocement tank	Blockwork tank	Plastic tank
Material costs (€)	470	500	870
Labour costs (€)	220	220	25
Total rwh Capital cost (€)	**690**	**720**	**895**

Operating Costs

There is no electrical pump fitted to the rwh system and therefore no electrical operating costs.

Maintenance Costs

There are no operating parts to maintain so maintenance costs are minimal. A nominal annual charge of €5 is assumed to cover any maintenance costs for cleaning of first flush or mesh filters.

Applying the economic assessment tools
Simple Payback

Table 8.11 summarises the respective payback periods for each tank option and each public supply option.

Table 8.11 Results for Simple Payback analysis for rwh system compared to alternative water supply options

Tank material	Communal supply	Water tap	Private vendor
Ferrocment	26.5 years	12.3 years	2.5 years
Blockwork	27.7 years	12.9 years	2.6 years
Plastic	34.4 years	16.0 years	3.2 years

Discussion

Table 8.9 shows the annual supply costs for House A (no rwh system) for each supply option as being €31, €61 and €285 respectively. Table 8.11 shows the corresponding payback periods by comparing each rwh tank option with the respective public water supply option. For example, the payback for the ferrocment rwh system when compared to the communal supply system will be the initial investment costs of €690 divided by the net annual savings which in this case is (€31 − €5) = €26 per year. The simple payback will be 26.5 years. The other rwh systems show paybacks of 27.7 and 34.4 years respectively when compared with the communal supply option. Each rwh option shows a payback of less than 3.2 years when compared to the private vendor supply option. When compared to the costs of purchasing water from a water tap the payback is of the order of 12.3 and 12.9 years for the ferrocement and blockwork tanks and 16 years for the more expensive plastic tank. We can now investigate further using the equivalent annual cost method.

Equivalent Annual Cost

Table 8.12 Annual Equivalent Cost Analysis for Freetown

		Ferrocment	Blockwork	Plastic
Financial Variables	Discount factor, r (%)	4%	4%	4%
	No. of years, n	20 years	20 years	20 years
	$PV_{annuity\ factor}\ \dfrac{1-(1+r)^{-n}}{r}$	13.59	13.59	13.59
Annual Equivalent Costs	Capital Costs	€ 690	€ 720	€ 895
	Annual Equivalent of Capital Costs	€ 51	€ 53	€ 66
	Total Annual Equivalent Cost (including maintenance & operating costs)	€56	€58	€71
	Total Present Value of rwh installation	€758	€788	€963
RWH production	Annual Harvestable Rainfall Yield, $Y_{R},a\ (m^3/yr)$	43.8 m³/yr	43.8 m³/yr	43.8 m³/yr
	Percentage Demand met (%)	100%	100%	100%
	RWH total yield over lifetime (m³)	876 m³	876 m³	876 m³
Volumetric Unit Costs	**RWH total unit production cost (€/m³)**	**1.27**	**1.32**	**1.62**
	RWH operation & maintenance unit production cost (€/m³)	**0.11**	**0.11**	**0.11**
	Communal Volumetric Tariff (€/m³)	**0.70**	**0.70**	**0.70**
	Water Tap Volumetric Tariff (€/m³)	**1.40**	**1.40**	**1.40**
	Water Vendor Volumetric Tariff (€/m³)	**6.50**	**6.50**	**6.50**

Discussion

To simplify the discussion we can now compare the unit costs for each supply option in Table 8.13. At a glance we can see that the rwh unit production costs for all tank options are significantly lower than purchasing water from a private vendor. They are also lower than purchasing water from a public water tap for the ferrocement and blockwork tanks. However, the unit production costs for the rwh system for all tank types are higher than the communal tariff.

Table 8.13 Unit costs for each water supply option

		Ferrocment	Blockwork	Plastic
Volumetric Unit Costs	RWH total unit production cost (€/m³)	1.27	1.32	1.62
	Communal Volumetric Tariff (€/m³)	0.70	0.70	0.70
	Water Tap Volumetric Tariff (€/m³)	1.40	1.40	1.40
	Water Vendor Volumetric Tariff (€/m³)	6.50	6.50	6.50

Chapter 9
Sustainable Development Goals (SDGs)

9.1 Introduction

Introducing the SDGs

The Sustainable Development Goals (SDGs) are a set of 17 goals which are designed to form a template for the world's future sustainable development through to 2030. The 17 goals are backed up by a set of 169 detailed targets. The goals were negotiated over a two-year period at the United Nations and were agreed to by nearly all of the world nations on 25th Sept 2015.

Although the SDGs are not legally binding, they are unique in that they are designed to provide a framework under which all countries are expected to take ownership and make choices now, to sustainably improve life for future generations. The SDGs are designed to recognize that ending poverty must go hand in hand with strategies that build economic growth and address a range of social needs including education, health, social protection and job opportunities, while tackling climate change and environmental protection. SDG 6 relates solely to ensuring availability and sustainable management of water and sanitation for all. However, water management underpins each of the other 16 goals as discussed later in this chapter.

© Springer Nature Switzerland AG 2021
L. McCarton et al., *The Worth of Water*,
https://doi.org/10.1007/978-3-030-50605-6_9

The SDGs were designed to form a template for Global Sustainable Development to 2030

9.2   Universality, Integration, Transformation

Many agencies misinterpret the concept of the SDGs and only focus on specific goals which they consider relevant to their own areas of expertise.

The Sustainable Development Goals are underpinned by three core concepts, Universality, Integration and Transformation.

The goals apply to every nation and every sector from cities, businesses, schools, community organization, all are challenged to act. This is called *Universality*.

Secondly it is recognized that the goals are all interconnected in a system. We cannot aim to achieve just one goal. We must achieve them all. This is called *Integration*.

And finally, it is widely recognized that achieving these goals involves making very big, fundamental changes in how we live on Earth. This is called *Transformation.*

Figure 9.1 presents a summary of each of the interrelated goals.

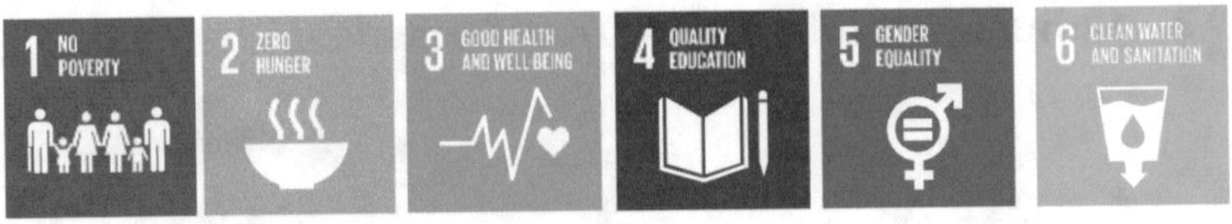

Goals 1 to 6 address the fundamental global development challenges of ;

- ending poverty in all its forms everywhere,
- end hunger, achieve food security and improved nutrition and promote sustainable agriculture
- ensure healthy lives and promote wellbeing for all at all ages
- ensure inclusive and equitable quality education and promote lifelong learning opportunities for all
- achieve gender equality and empower women and girls
- ensure availability and sustainable management of water and sanitation for all

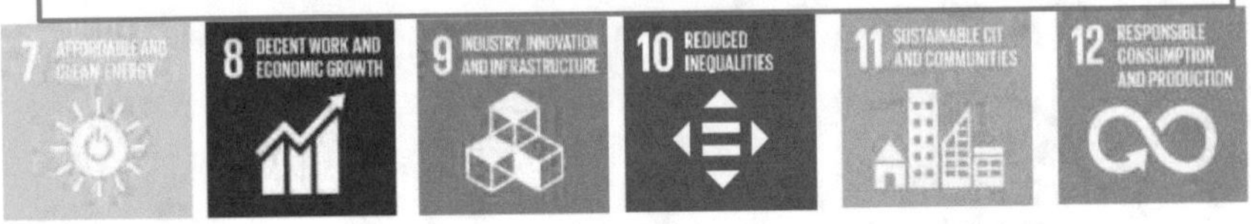

Goals 7 to 12 address global climate and sustainability challenges to ;

- Ensure access to affordable, reliable, sustainable and modern energy for all
- Promote sustained, inclusive, and sustainable economic growth, full and productive employment and decent work for all
- Build resilient infrastructure, promote inclusive and sustainable industrialization, and foster innovation
- Reduce inequality within and among countries
- Make cities and human settlements inclusive, safe, resilient and sustainable
- Ensure sustainable consumption and production patterns

Goals 13 to 17 address challenges to;

- Take urgent action to combat climate change and its impacts
- Conserve and sustainable use the oceans, seas and marine resources for sustainable development
- Protect, restore and promote sustainable use of terrestrial ecosystems, sustainable manage forests, combat desertification, halt and reverse land degradation, and halt biodiversity loss
- Promote peaceful and inclusive societies for sustainable development, provide access to justice for all, and build effective, accountable, and inclusive institutions at all levels
- Strengthen the means of implementation and revitalize the global partnership for sustainable development

Fig. 9.1 Making Sense of the SDGs

9.3   Water and the 2030 Sustainable Development Goals

Fig. 9.2 Water is fundamental to the achievement of each of the 2030 Sustainable Development Goals

The importance of water in each of the SDGs is summarised below.

SDG 01 NO POVERTY. Many people in developing countries continue the daily struggle to access a sufficient quantity of safe drinking water. They manage on less water per capita per day than those in developed countries and typically pay more of their income to access that water. Coupled with environmental pollution are high malnutrition rates and lack of basic sanitation services. Water poverty continues to undermine success in other goals and contribute to keeping a huge percentage of the global population in the poverty trap.

SDG 02 ZERO HUNGER. All food production depends on a reliable supply of water, either from surface or groundwater. Stresses on global water sources means a stress on the ability of the global population to feed itself. Water is the largest consumer of the world's freshwater resources. The western diet requires 3500 litres per day to produce. By 2050 a 19% increase in agricultural water consumption is expected. Food insecurity and high levels of malnutrition continue to disportionately effect parts of the world and is a major factor in high child mortality rates.

SDG 03 GOOD HEALTH AND WELL-BEING. Many water related diseases such as cholera, typhoid and dysentery are caused by lack of sanitation, hygiene and safe drinking water. Other diseases are connected to the hydrological system because vectors that transmit them breed in water courses and standing water. Malaria and dengue fever are the most deadly with 214 million cases of malaria annually, many of them in Africa.

SDG 04 QUALITY EDUCATION. Water poverty directly impacts on access to education particularly in water stressed regions. The task of fetching water typically falls to children, particularly females and prevents many accessing a basic education. 31% of schools lack access to safe water and adequate sanitation globally. Children with high rates of malnutrition lack ability to focus and continue their education.

SDG 05 GENDER EQUALITY. In Sub-Saharan Africa 75% of people have no household connection and walk many kilometres each day to fetch water. This task mostly falls to females. This means that young girls often fail to attend school due to the daily burden of water access. Globally, 1 in 4 girls do not complete primary school compared to 1 in 7 boys.

SDG 06 CLEAN WATER AND SANITATION. The Joint Monitoring Programme Report 2019, *Progress on drinking water, sanitation and hygiene: 2000–2017,* estimates that 1 in 3 people globally do not have access to safe drinking water and more than half the world does not have access to safe sanitation services. This equates to 2.2 billion people around the world who do not have access to a safe drinking water source, 4.2 billion people without access to safely managed sanitation services and 3 billion lacking basic hand washing facilities.

SDG 07 AFFORDABLE AND CLEAN ENERGY. The water – food – energy nexus is a core component of sustainable global development. Demand for all three is increased by stresses caused by rising global population, rapid urbanization, and intensification of agricultural production causing change in land use.

SDG 08 DECENT WORK AND ECONOMIC GROWTH. Investments in water security and economic growth are interlinked. Agriculture accounts for 70 per cent of water extractions globally. The economic risks from droughts and floods are most pronounced in agriculture dependent economies. The greatest economic losses are from inadequate water supply and sanitation and are highest in developing countries. This inhibits their economic growth potential. The economic cost of climate change on water resources has not been fully quantified but is estimated to be in the region of $120 billion per year from urban property flood damage alone.

SDG 09 INDUSTRY, INNOVATION AND INFRASTRUCTURE. A World Bank Report in 2016 estimated the costs of meeting the 2030 SDG targets for drinking water, sanitation and hygiene at 1.7 trillion USD. This would require a threefold increase on current global investment. This does not include the cost of operating this new infrastructure.

**SDG   10
R E D U C E D
INEQUALITIES.**
Water   poverty
remains a drag on
many communities
throughout   the
world. Access to
safe drinking water and sanitation, a reliable supply of water for food production
and industrial production remains a huge
challenge for many countries.

**SDG   11
SUSTAINABLE
CITIES   AND
C O M M U N I T -
IES.** More than
half of the world's
population now live
in   cities.   Much
existing infrastructure is already ineffi-
cient. Up to one third of potable water pro-
duced is lost through leakage in aging water supply networks. Cities are also refuges
for those fleeing rural poverty or war. Many end up living in areas with no water or
sanitation services in informal settlements known as slums or shanty towns.
Globally, 1 in 3 people in urban areas live in such a slum household.

SDG 12 RESPONSIBLE CONSUMPTION AND PRODUCTION. Industrialised countries consume water embedded in the production of goods and services. The proportion of water used in a production process is called the "Water Footprint". Food production is a key component of this footprint. A typical western diet high in meat and dairy requires 3500 litres per day to produce. To produce 1 kg of beef requires 15,415 litres of water. Contrasting this with small farmers in Africa who lack basic technology to produce food beyond a subsistence level.

SDG 13 CLIMATE ACTION. Water plays a vital role in the health of our environment. Climate change interacts with the hydrological environment through melting glaciers, rising sea levels, rising temperatures and a host of other interrelated ecosystems. Flooding is becoming more frequent. As temperatures rise, evapotranspiration from land, seas and plants increases the rate of desertification with increased frequency of droughts. Changing rainfall patterns has an impact on river basin networks, affecting water quantities and qualities. All of these interrelated effects are likely to impact on the ability of entire ecosystems which support human life to survive.

SDG 14 LIFE BELOW WATER. Rises in sea level as a result of the effect of increased rates of global warming on oceans will have a direct impact on low lying islands, deltas and coasts. The temperature in Antarctica's ice-sheet has risen 10 times faster than the global average. This has resulted in changes in glaciers, ice-sheets, permafrost, snow melt volumes and will impact on ocean flows in complex ways.

SDG 15 LIFE ON LAND. Changes in the ecological characteristics of water ecosystems can impact all living organisms that inhabit it. Habitat change within water systems impacts on species such as water birds, wetland dependent mammals, freshwater fish, freshwater amphibians.

SDG 16 PEACE, JUSTICE AND STRONG INSTITUTIONS. Ancient civilisations recognised that equitable sharing of water resources was intrinsic to sustainable development of their society. With demand for water globally expected to increase by 40–50%, improved water resource management on regional and even global levels will be required.

SDG 17 PARTNERSHIP FOR THE GOALS. The importance of partnerships in achieving the SDGs is recognised by giving it a goal in itself. Closing the water gap will also require partnerships as more than 260 river basins are shared between countries.

9.4   Making Sense of the SDG's

The starting point of any discussion of the SDGs is firstly to address the concept of "sustainability". There are a number of generally agreed definitions;

"Sustainability" The possibility that human and other forms of life on earth will flourish forever …John Ehrenfield, Professor Emeritus, MIT.

"Sustainable Development" meeting the needs of the present generation without compromising the ability of future generations to meet their own needs …. Bruntland Commission, 1987.

"Sustainable Development" Enough – for all – forever …African Delegate to Johannesburg (Rio +10).

9.5   Enough – For all – Forever

The SDGs were proposed as a template to achieve the unending concept of sustainable global development for all. They are based on the original concepts of *"sustainable development"*. These definitions were postulated on the assumption that the Earth's natural resource base would facilitate developing countries to catch up with living standards in developed countries, facilitate continued growth in high income countries, and support a sustainable global population growth. To catch up with richer countries, and achieve a more equitable lifestyle for their peoples, developing countries could invest in technologies, infrastructure and education and step by step narrow the income and poverty gaps. Yet clear evidence from the Intergovernmental Panel on Climate Change (IPCC) has detailed that the Earth's resources are not finite and that the development of high income economies based on fossil fuels has

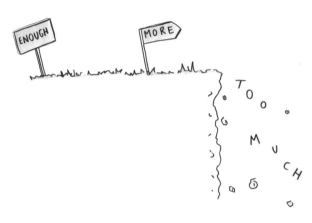

resulted in the current "climate emergency", where we are forced to make definite decisions which will affect future generations ability to survive and thrive. This leads us to question the very premise upon which the SDGs were proposed, *"enough – for all – forever"*.

9.6   The Contradiction of Growth

We now have robust scientific evidence that shows the effects of human induced changes on how the Earth's system oper- ates, from melting of the ice sheets, to chang- ing rainfall profiles to loss of ecosystems and biodiversity. The goals of ending all poverty in all its forms every- where, together with the need to create an economic and human  development template that will function within the boundaries of the overstretched life-support systems on earth form seemingly unachievable goals within the current SDG framework.

9.7   End Poverty in All Its Forms Everywhere

 One of the major goals of the SDGs is to eradicate poverty in all its forms by 2030. The original mea- sure of extreme poverty proposed in 2015 was the number of people living on less than $1.25 a day. The world bank has since revised this measure to $1.90 per day for extreme poverty. However, many studies suggest that this measure is not sufficient to provide basic nutrition and other requirements for sur- vival, including food, safe drinking water, shelter and

clothing. If this measure is too low to secure a fair chance of surviving the first year of life why are we still using it? In order to achieve this reduction in poverty, the SDGs propose a target for sustained economic growth of at least 7 per cent gross domestic product growth per annum in the least developed countries. **Even if this goal were possible, it is driven by the current unsustainable economic model of increased use of the earth's resources based on a largely fossil fuel economy.** This is one of the main criticisms of the SDGs, that they fail to propose a more sustainable means to achieve their stated goal of ending poverty which can create a more equitable society and symbiosis with the natural world.

9.8   Are the SDGs Ideals Masquerading as Goals?

A strength of the SDG 2030 agenda is that the SDGs recognize that ending poverty must go hand in hand with strategies that build economic growth and address a range of social needs including education, health, social protection and job opportunities, while tackling climate change and environmental protection. However as with any target, the SDG performance goals should be specific, measurable, achievable, relevant and time bound. One of the harshest critics of the SDGs has been former world bank economist, William Easterby. In a 2015 Economist article he criticised the SDGs as being unactionable and unquantifiable, unmeasurable and unattainable and that they were merely *"Ideals masquerading as targets"*. In an open letter to the UN signed by Noam Chomsky, Naomi Klein, Chris Hedges and other powerful voices, they stated that *"as currently formulated the SDGs merely distract us from addressing the challenges we face"*. They also highlighted in their correspondence that the SDGs evade the burden of debt which underpins many countries inability to rise up the economic ladder. The SDGs risk becoming a marketing tool. "SDG washing" is an evolution of the term "green washing" originally used to describe companies that portray themselves as environmentally friendly when in reality they are not. "SDG washing" refers to companies that use the SDGs as "window dressing" to present a deceptively positive picture of their environmental and social impacts. Aligned with the SDG 2030 agenda is the concept of planetary boundaries as discussed in Chap. 11. The concept of planetary

IS IT A GOAL IF IT'S UNATTAINABLE?

boundaries has been developed to outline a safe operating space for humanity that carries a low likelihood of harming life support systems on Earth to such an extent that they are no longer able to support economic growth and human development. Is the SDG framework for global sustainable development compatible within the planetary boundaries? Is a new revision of the SDGs required which proposes sustainable development goals which can be achieved within the planetary boundaries, in which rich countries reduce their consumption levels to allow developing countries grow until they converge at the lower income of high-income countries? Or will a business as usual approach continue where rich countries scramble for scarce resources with lower income countries pushed out of the marketplace and failing to develop. Within this context, we need to revise the concepts of the SDGs and propose a new sustainable development agenda with achievable goals and targets which allow countries to thrive within the planetary boundaries.

Chapter 10
Resilient Cities and Communities

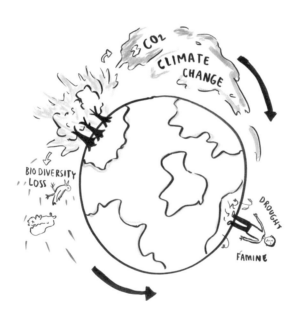

Fig. 10.1 Building resilient cities and communities will be increasingly difficult given the challenges of climate change and urbanisation

10.1   Introduction: Build Solid Ground Project

Build Solid Ground is a project implemented by a consortium of 14 organisations in seven EU countries. Solid Ground is a global advocacy campaign managed by Habitat for Humanity Europe, Middle-East & Africa (EMEA) which focuses on Sustainable Development Goal (SDG) 11. It aims to educate Europeans on current urban and housing challenges and inspire city authorities, governments and citizens to work together to find solutions to urban problems in both developed and developing countries (www.habitat.org/build-solid-ground). Figure 10.1 illustrates the challenges of climate change and urbanisation.

This project is funded by the European Union

10.2 Understanding the Components of a Resilient City and Community

01 VULNERABILITY

WHAT IS VULNERABILITY?

Vulnerability comes from the Latinword for"wound,"vulnus, and may be defined as *being susceptible to being damaged or harmed.*

In terms of cities, vulnerability is used to denote weaknesses that expose the city to harm.

This harm can be inflicted on the infrastructure, the population or the food supply.

These weak/vulnerable points can be due to climate change, the location of the city, poor planning and law enforcement or poor infrastructure.

02 RESILIENCE

WHAT IS RESILIENCE?

Resilience comes from the Latin word resilio, which means to spring back, and may be defined as *the ability of a system or city to cope with disturbing or destabilising forces.*

These forces may be climate change, natural disasters or human induced events.

Therefore resilient cities are designed and built to cope with disturbance and destabilisation and to absorb these events and continue as near to normal as possible.

03 | DIVERSITY

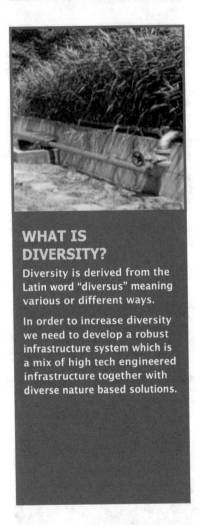

WHAT IS DIVERSITY?

Diversity is derived from the Latin word "diversus" meaning various or different ways.

In order to increase diversity we need to develop a robust infrastructure system which is a mix of high tech engineered infrastructure together with diverse nature based solutions.

04 | SUSTAINABILITY

WHAT IS SUSTAINABILITY?

Sustainability involves the long term maintenance of well-being, which in turn depends on the well-being of the natural world and the responsible use of natural resources.

Therefore we can reduce vulnerability in cities by sustainably promoting resilience, i.e. improving the ability of the city to resist disturbance in a resource efficient manner.

The components to understand in developing a concept for a resilient city and community are the concepts of vulnerability, resilience, diversity and sustainability. The characteristics of developing a sustainable city and community, which are to reduce Vulnerability, increase Resilience and increase Diversity.

Fig. 10.2 Shows the requirements to achieve a sustainable resilient city and community

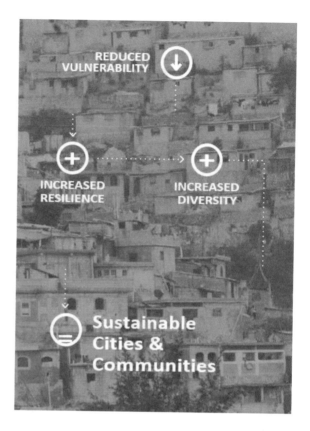

10.3   SDG11: *"Make cities and human settlements inclusive, safe, resilient and sustainable"*

In 2015 almost 4 billion people (54% of the world's population) lived in cities and that number is projected to increase to 5 billion people by 2030. Almost three quarters of the EU population already live in urban areas, cities and towns with over 40% residing in cities alone. The urban population of Europe is projected to rise to over 80% by 2050. Such rapid urbanization brings enormous challenges for basic infrastructure services and combined with unplanned urban sprawl, makes cities and communities more vulnerable to disasters. If we are going to manage as a human race, we need to make sure cities work for us. They have been seen as attractive places for jobs and opportunities, and that means that assets and people are concentrated – so, if there is an event like an earthquake or a storm, the losses can be huge. SDG11 is focused around a set of targets that seek to make the world's human settlements and urban spaces more inclusive, safe, resilient and sustainable (Fig. 10.3).

Fig. 10.3 Shows the targets of SDG11 to make cities and human settlements inclusive, safe, resilient and sustainable

Building safety into cities is key, particularly as the world faces an uncertain future with climate change. Jo da Silva, the founder and leader of ARUP International Development which works with organizations committed to improving human development outcomes says that "Cities will always experience shocks and stresses, that is normal – but the things that make cities work are the infrastructure such as water supplies and energy. Put simply, a hospital isn't just a building, it's a critical asset within the healthcare system that serves society. We need to think about the social outcomes of our projects, not just physical outputs. It's important that we build cities that are as resilient as they can be." This focus on resilience and the urban/peri urban (i.e. city outskirts) environment is also a focus of the Sustainable Development Goals, in particular SDG 11.

Chapter 11
Planetary Boundaries

© Springer Nature Switzerland AG 2021
L. McCarton et al., *The Worth of Water*,
https://doi.org/10.1007/978-3-030-50605-6_11

Thigh bone connected to the hip bone
Hip bone connected to the backbone
Back bone connected to the shoulder bone
Shoulder bone connected to the neck bone
Neck bone connected to the head bone

James and Rosamond Johnson.

11.1 Think Global Act Local

The slogan *"think global act local"* is a favourite of environmentalists. It is used to express in a simple and concise manner the fact that the global environmental issue of climate change is due to the actions of each one of us individuals at a local level. While the slogan is applied and used in relation to climate change, it is applicable to the whole crisis facing the planet. However, the breadth and connectedness of the wider global problem is not easy to grasp.

It is complex, multidimensional and difficult to condense into "bite sized" language. Explanations of the current global environmental crisis tend to use scientific language, and concepts, that are not generally used in everyday conversations. This terminology requires an understanding of language, prior to allowing an understanding of the concepts being discussed. It is therefore, to a certain extent, counterproductive, in that terms and concepts that are meant to explain the current environmental crisis end up becoming barriers to an understanding and comprehension of this crisis. As a result, there tends to be a focus on climate change at the expense of an understanding of the other global environmental threats.

The interconnectedness of the Earth's systems, and how they affect each other, is not generally grasped. The fact that the systems of planet Earth which control how we live on this planet are poorly understood also has profound effects on how we act local but *think* global. The effects of our actions locally, must be seen in their entirety, before meaningful changes can be made to our way of life such that global changes will occur.

Taking one planetary system in isolation, water, it is manifest at a local and global level we are in an age of crisis. At a local level there are problems in urban shanty towns where residents face no piped water or sanitation facilities. Others face droughts and floods which force them to leave where they live resulting in the people ending up as water refugees. Globally the increased population requires increased agricultural production thus using more freshwater resources. This has resulted in the depletion of aquifers and lowering of groundwater levels. Increased industrial output has also resulted in contamination of natural water sources. Environmental degradation and biodiversity have been impacted by removal of wetlands, dam construction and urbanisation. Other global factors include melting of glaciers, the rise in sea levels, and desertification. We can see that to fully understand the local problems relating to water, the discussion has now moved outside of the topic of water resources and enters other areas such as population levels, methods of food production, reduction of wildlife and interference with nature, rising sea levels, land use etc. Therefore, complexity and interconnectedness return to a discussion that started off dealing with a single issue, local water resources.

Many attempts have been made to promote a concept or framework that includes all the environmental threats facing the planet. These include the Millennium Development Goals (MDG's) and The Sustainable Development Goals (SDG'S). However possibly the most useful, easy to use and least cumbersome framework is that proposed and developed by the Stockholm Resilience Centre at Stockholm University (www.stockholmresilience.org). This is called the "**Planetary Boundaries**" system.

11.2 Planetary Boundaries

First published in 2009, this was one of the first attempts to undertake an integrated assessment of all the major environmental threats to the sustainability of humanity and the Earths ecosystems. **It is based on science and it identifies nine processes and systems that regulate the stability and resilience of the Earth system** – the interactions of land, ocean, atmosphere and life that together provide conditions upon which our societies depend. The planetary boundaries (PB's) framework arises from the scientific evidence that the Earth is a single, complex, integrated system— that is, the boundaries operate as an interdependent set. The Earth operates in a manner in which these processes and their interactions can create stabilizing or destabilizing feed backs. **This has profound implications for global sustainability, because it emphasizes the need to address multiple interacting environmental processes simultaneously** (e.g., stabilizing the climate system requires sustainable forest management and stable ocean ecosystems). In doing this it proposes a framework whereby the complexity of the planet and the interconnectedness of the Earth's systems can be dealt with in a useful, understandable and comprehensive way. The PB's propose that each of these systems and processes, should stay within fixed measurable boundaries to decrease the risk of irreversible and potentially catastrophic shifts in the Earth system. **These boundaries define the upper limit of environmental effects at the global scale.** PB's are global targets and this framework relates to Earth system processes at planetary scales. It is a practical method for considering multiple anthropogenic, global, and environmental pressures. Transgressing a boundary increases the risk that human activities will drive the Earth system into an unstable state damaging efforts to reduce poverty and leading to a deterioration of human well-being in many parts of the world, including wealthy countries. To date, eight of the nine planetary boundaries have the boundary quantified. **Four of the planetary boundaries have now been crossed as a result of human activity and two of these, climate change and biodiversity loss, are regarded as "core boundaries". <u>A core boundary is referred to as one that when it is breached the resultant effect would be to drive the Earth system into a new state.</u>** As the Earth system moves beyond planetary boundaries and into areas of increasing risk, ecosystems may change rapidly and dramatically, affecting ocean acidification, eutrophication, and ambient temperatures. Ensuing changes will pose threats to agricultural production, infrastructure, and human health.

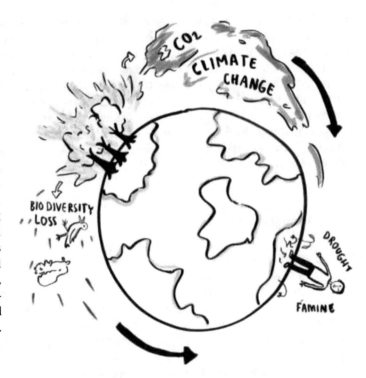

The Nine Planetary Boundaries are:

- **Climate Change**
- **Biodiversity loss and species extinction**
- **Stratospheric Ozone Depletion**
- **Ocean Acidification**
- **The Phosphorus and Nitrogen Cycle (Biogeochemical flows).**
- **Land-system change (e.g. deforestation).**
- **Freshwater use.**
- **Atmospheric Aerosol Loading**
- **Chemical Pollution (e.g. organic pollutants, radioactive materials and plastics).**

11.3   Climate Change

The United Nations Framework Convention on Climate Change (UNFCCC), defines climate change as a *"change of climate that is attributed directly or indirectly to human activity that alters the composition of the global atmosphere and that is in addition to natural climate variability observed over comparable time periods"*. Warming of the climate system is unequivocal, as is now evident from observations of increases in global average air and ocean temperatures, widespread melting of snow and ice and rising global average sea level. **The Planetary Boundary for climate change was set at a CO_2 concentration in the atmosphere limited to 350 ppm and/or a maximum change of +1 W m^{-2} in radiative forcing**. This boundary was passed a number of years ago and the present value for CO_2 levels is close to 407 ppm. Humanity enters a danger zone when the concentration of CO_2 in the atmosphere reaches a range between 350 and 550 ppm, compared with a preindustrial level of 280 ppm. A safe temperature increase should be limited to 2 °C. The goal of a number of climate change measures is 450 ppm, however it is not clear that this figure would guarantee the 2 °C change predicted.

The chances of limiting global temperature rises to this figure are now considered unrealistic. A temperature increase of several degrees will have many consequences as the interconnections between greenhouse gas concentrations are not linear. Melting glaciers will result in more heat being absorbed by the earth mostly due to changes in albedo (the reflection of light from the earth's surface). Weather patterns will change with precipitation becoming more variable and increasing in many areas of the globe while arid areas are likely to become even dryer. As a result of extreme weather more storms and heat waves will occur, and this will impact on food production while water availability will be reduced in many areas. Sea level rise will have a major impact on coastal areas.

11.4   Biodiversity Loss and Species Extinction

<u>Biodiversity refers to the variety of life on Earth.</u> There are three main types of biodiversity. These are **genetic diversity** which is to do with differences in DNA among individuals, **species diversity** which deals with the variety of species in a given area and **ecosystem diversity** which has to do with the variety of habitats, communities and ecosystems. The greater the variety of species the healthier the biosphere (bio meaning living). **It is biodiversity that maintains the health of the Earth, its people and its flora and fauna.** It does this by promoting a variety of genes, species and ecosystems in order to promote resilience and the ability to withstand threats. Biodiversity is responsible for our food, our medicines and ultimately our economy. However, biodiversity is not evenly distributed across the globe. The diversity of all living things (the biota) depends on a range of parameters including temperature, precipitation, altitude, soils, geography and the presence of a range of different species. Humans have been the cause of biodiversity losses historically as well as in the present. Losses have tended to follow colonisation. **Biodiversity is another planetary boundary of major concern that has been passed.** The specific boundary chosen as a measure for biodiversity is the rate of extinction. Before industrialization, the extinction rate was less than one species per million species each year. At present more than 100 species per million are going extinct each year. The proposed boundary is set at 10 species per million species per year. Once extinct, a species can never return. When extinction rates surpass the normal background rate, this is referred to as **mass extinction**. These have occurred at least five times in the Earth's history with between 25–50% of species lost. Current extinction rates are 100–1000 times greater than the usual background rate. Presently the world contains two thirds of the animal population compared to 40 years ago. The species abundance of wild animals (the number of individuals per species) has also dropped by 40% in the same time. This has led to the current situation being called the **6th mass extinction**. This has occurred in the last thirty years. Biodiversity is referred to as the control panel for those living on Earth. Changes in biodiversity will have major effects on the Earth system. Biodiversity is not only about species numbers it concerns variability in terms of habitats, ecosystems, and biomes.

11.5   Stratospheric Ozone Depletion

The stratospheric ozone layer in the atmosphere filters out ultraviolet (UV) radiation from the sun. If this layer decreases, increasing amounts of ultraviolet radiation will reach ground level. This can cause a higher incidence of skin cancer in humans as well as damage to terrestrial and marine biological systems. The appearance of the Antarctic ozone hole was proof that increased concentrations of ozone-depleting chemical substances, interacting with polar stratospheric clouds, had passed a threshold and moved the Antarctic stratosphere into a new regime. Fortunately, because of the actions taken as a result of the Montreal Protocol, we appear to be on the path that will allow us to stay within this boundary. The cooling of the stratosphere is a result of the atmosphere being warmed by human induced greenhouse gas emissions. What we at the planet's surface call global warming, is experienced high above as cooling, as the normally outgoing heat from the earth is retained at the lower level instead of warming the stratosphere. This trapped heat causes the cooling and also increased formation of ozone depleting substances. There is a considerable lag time between decreases of greenhouse gases (GHG's) in the troposphere and stratosphere ozone recovery. Though the situation is improving the Artic and Antarctic ozone holes may continue for a decade or two. However, it is a positive story of human action resulting in the current position of us being within this planet boundary.

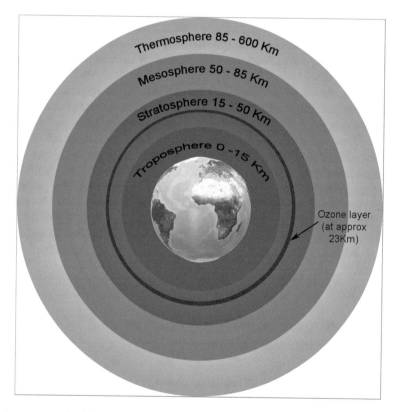

Fig. 11.1 Stratospheric layers

11.6   Ocean Acidification

Increasing concentrations of Carbon dioxide in the atmosphere have increased the surface acidity of the world's oceans. This is due to the fact that carbon dioxide dissolves in seawater, reacting with water to form Carbonic Acid.

$$CO_2 + H_2O ------- > H_2CO_3$$

Ocean acidification threatens to push marine life over catastrophic tipping points. There is no other source of increased CO_2 in the atmosphere than human emissions from burning fossil fuel energy sources and degradation of ecosystems. Around a quarter of the CO_2 humanity emits into the atmosphere is ultimately dissolved in the oceans. Here it forms carbonic acid, altering ocean chemistry and decreasing the pH of the surface water. This increased acidity reduces the amount of available carbonate ions, an essential 'building block' used by many marine species for shell and skeleton formation. Beyond a threshold concentration, this rising acidity makes it hard for organisms such as corals and some shellfish and plankton species to grow. Losses of these species would change the structure and dynamics of ocean ecosystems and could potentially lead to drastic reductions in fish stocks. Compared to pre-industrial times, surface ocean acidity has already increased by 30 percent.

Unlike most other human impacts on the marine environment, which are often local in scale, the ocean acidification boundary has global impacts. It is also an example of how tightly interconnected the boundaries are, since atmospheric CO_2 concentration is the underlying controlling variable for both the climate and the ocean acidification boundaries, although they are defined in terms of different Earth system thresholds. A monitoring point or planet boundary proposed is that the oceanic aragonite saturation state is maintained at least at 80% of 3.44 ppm, which is the average global preindustrial seawater saturation state. Aragonite is proposed as it is the type of calcium carbonate most sensitive to acidification. It is also the kind that reef building corals precipitate when building their skeletons. Where the aragonite saturation state falls below 1ppm, aragonite starts to dissolve. Therefore, the planetary boundary was set well away from the aragonite saturation state at dissolution.

11.7   The Phosphorus and Nitrogen Cycle (Biogeochemical Flows)

Nutrient overload of nitrogen (N) and phosphorus (P) in groundwater, lakes, rivers and estuaries are regarded as local problems. However, when these local problems are combined, they become a global concern. Human activities now convert more nitrogen gas (N_2) from the atmosphere into reactive forms (ammonium, nitrate) than all of the Earth's terrestrial processes combined. The current flow of phosphorus to the oceans is about three times the pre-industrial flow, driven by the mining of phosphorus and the transformation of it into reactive forms primarily for agriculture. As a result, the global P and N cycles are now driven at a global level by the actions of humans. N and P are the two macronutrients that support our ability to feed the world. However, nutrient loads are causing abrupt changes in aquatic ecosystems causing them to tip over thresholds. The resilience of these systems is being eroded. 17% of the Baltic Sea is a dead zone due to decades of run off from Sweden, Finland, Denmark, Russia Germany Poland and the other Baltic States. As a result, the Baltic Sea has increased growth of planktonic algae which consume oxygen as they die and decay. Similar dead zones exist in seas in the Gulf of Mexico, New Zealand and China. Eutrophication, which is defined as an excess of nutrients in an aquatic system, ranges from 54% of all freshwater lakes and reservoirs in Asia to 53% and 48% in Europe and North America respectively to 41% in Latin America and 28% in Africa. The primary purpose of most reactive N is to enhance food production through fertilisation. The conversion of Nitrogen (N) takes place primarily through four processes:

- Industrial fixation of atmospheric nitrogen gas, (N_2) to ammonia (approx. 80 Mt N per annum),
- Agricultural fixation of atmospheric nitrogen gas, (N_2) through the cultivation of leguminous crops (approx. 40 Mt N per annum)
- Fossil fuel combustion (approx. 20 Mt N per annum) and
- Biomass burning (approx. 10 Mt N per annum).

Nitrous oxide is included in the climate change boundary as it is one of the most important greenhouse gases. Although they are separate processes the N and P cycles are included in one boundary as they drive abrupt shifts in Earth's subsystems as they are key biological nutrients. They both affect human life support systems, and both have significant impacts on a planetary scale. The total amount of P that leaches from land and is transported by freshwater systems to the sea amounts to approx. 22 million Mt of P per annum. **The current scientific assessment is that the safe planetary boundary for nitrogen is a global extraction of N_2 from the atmosphere amounting to 25% of the current extraction up to 35 Mt per annum.** Unlike nitrogen, phosphorus is a finite fossil fuel that is both mined for human use and added to the Earth system through weathering. Some experts have warned that we may lack enough phosphorus to meet future demands for crop protection and may indeed be at peak phosphorus. There are two considerations for a P planet boundary to consider: Phosphorus induced anoxic events in the oceans that could trigger global regime shifts in the oceans and phosphorus induced collapses of aquatic ecosystems due to phosphorus overload in freshwater ecosystems. **A suggestion is setting the safe planetary boundary at keeping the inflow of P to the oceans at lower than 10 times the flows of P before modern agriculture.** However, this means that the planetary boundaries to freshwater systems has been transgressed while the one for anoxic events looms in the future. Therefore, with the world population expected to be 9 billion in 2050 feeding the world will become critical and so will phosphorus use. The challenges are;

- To avoid an ecological tipping point driven by phosphorus;
- To meet the need for phosphorus for food production in different parts of the world;
- To address the risk of "peak Phosphorus" at a time when P demand grows with food demand.

The key to P seems to be conserving P in agricultural ecosystems and to close P cycles in modern societies by returning P from cities to crop land. The effects of human activities on the phosphorus cycle are threefold.

- Large quantities of phosphorus (a finite element) are mined to produce fertilisers and detergents resulting in a depletion of soil resources,
- Phosphorus in tropical soils is removed by clearing forests where the phosphate is washed away in heavy rains, and
- Run off from fertilisers and animal wastes together with discharge of municipal sewage are added in excess to watercourses resulting in dissolved oxygen conditions and eutrophication.

The human impacts on the nitrogen cycle in the last 100 years include adding nitric oxide and nitrous oxide to the atmosphere, removal of nitrogen from topsoil, addition of nitrogen compounds to water courses leading to deterioration of water quality.

The usage of P is still within the boundary. However, the amount of N taken out from the atmosphere is beyond what would be considered a sustainable level. The present amount of atmospheric N that is removed is 121 million tons per year; the proposed boundary is set at 35 million tons per year. Thus, this is a boundary that has been passed by a wide margin.

11.8   Land-System Change (e.g. Deforestation)

Land-use change, primarily driven by intensification of agriculture, is one driving force behind the serious reductions in biodiversity, and it has impacts on water flows and on the biogeochemical cycling of carbon, nitrogen and phosphorus and other important elements. While each incident of land cover change occurs on a local scale, the aggregated impacts can have consequences for Earth system processes on a global scale. A major challenge with setting a land use boundary is that it needs to reflect not just the absolute quantity of unconverted and converted land but also its function, quality and spatial distribution. Agricultural expansion has been at approximately 0.5–1% per annum, or 25–50 million hectares per annum. To safeguard a stable climate, avoid losing vital species and secure freshwater flows a safe planetary boundary must be defined for land. **This will involve an end to the global expansion of agriculture**. All nations must accept responsibility for preserving the remaining rainforests as they are the most important global sinks and the site of terrestrial biodiversity.

The current land use types must be sustained in all their diversity, i.e. grasslands, wetlands, savannahs, steppe areas, shrub lands. Agriculture must be transformed by improving soil productivity i.e. it's capacity to hold and release nutrients for crops. Agriculture must also be transformed by converting farmland from being sources of carbon to becoming **carbon sinks**. Approximately 25% of global CO_2 emissions are sequestered on land. Land is thus a CO_2 sink. However, land use currently releases approximately 17% of global CO_2 emissions due to burning of carbon when soils are ploughed and losses from unsustainable agricultural practices. Deforestation releases approximately 18%, as forest is converted to cropland. The planet is a potential carbon sink on land, despite massive losses of CO_2 from agricultural land. The planet boundary set for land use is that no more than 15% of the global ice-free land surface be converted to cropland. Currently 12% of the global land surface is under crop cultivation. The remaining 3% will be reached in the coming decades and might include abandoned cropland in Europe, North America, Russia, Latin America and Africa's savannah. Cropland should be allocated to the most productive areas and processes that lead to the loss of productive land i.e. land degradation, loss of irrigation water, urban development and biofuel production should be controlled. Demand processes will also require management. Included here are diet, per capita food consumption, wastage in the food distribution chain. The land use boundary would require implementation at various levels through a global land architecture.

11.9   Freshwater Use

Green water is the amount of rainfall that is either intercepted by the vegetation, or enters the soil and is picked up by plants and evapotranspired back into the atmosphere. **Blue water** is the freshwater: surface and groundwater. It is stored in lakes, streams groundwater, glaciers and snow. The loss of green water due to land degradation and deforestation threatens terrestrial biomass production and carbon sequestration. Shifts in the volumes and patterns of blue water run off threaten supplies of blue water for sustainable aquatic ecosystems. The decline in moisture feedback vapour flows (green water flows) affects local and regional rainfall patterns. Key links between the planetary boundaries of water, land, biodiversity and climate are the green water flows or the moisture feedbacks. Deforestation alone has decreased green water flows by approx. 3000 km³ per annum, which is about the same volume as the total global use of water for irrigation. This is of great importance given that in many regions of the world more than 80% of rainfall comes from local to regional return flows of the moisture feedback. 80% of China's rainfall depends on land use in Eurasian countries like Pakistan, Iran, Iraq, Nepal and Russia. Rainfall over the Congo basin, the most water rich region in Africa, comes from evaporation in Eastern Africa. The Congo basin in turn provides rainfall to the Sahel region. To set a safe planetary boundary for freshwater resources three issues must be addressed.

1. The boundary must be set allowing enough green water flows to regenerate precipitation.
2. It must allow for such terrestrial ecosystem functions as carbon sequestration, biomass growth, food production and biological diversity.
3. It must secure the availability of blue water resources for aquatic ecosystems.

It is difficult to define a single freshwater boundary that captures the complexity of the water cycle. Consumptive water is the amount of blue water use and is an aggregate indicator of the amount of water used in rivers. By transgressing a boundary of 4000 km^3 per annum of consumptive blue water use the risk is that green and blue water induced thresholds, at regional and continental scales, will be approached. This would result in major changes in moisture feedback and freshwater/ocean mixing. Consumptive water use is currently estimated at 2600 km^3 per annum.

Global freshwater problems include the uneven distribution of water resources. As a result, though the quantity of freshwater on the planet can meet human needs, some areas of the world experience water scarcity conditions. Other problems include climate change, demands for drinking water as many developing countries have insufficient water to meet potable water and household needs, world population growth and the fact that water resources form boundaries between nations and therefore ownership (and sharing) of resources are contentious issues. The increase in global population means an increase in freshwater requirements thus limiting the available quantities of drinking water available. This in turn limits the quantity of water available for agricultural food production which is needed to feed the increased population. Reduced water levels in rivers streams and lakes results in slow water flow and the exhaustion of underground aquifers. Poor irrigation causes much of the water to be evaporated before it can be used.

11.10 Atmospheric Aerosol Loading (Microscopic Particles in the Atmosphere That Affect Climate and Living Organisms)

An atmospheric aerosol planetary boundary was proposed primarily because of the influence of aerosols on Earth's climate system. Through their interaction with water vapour, aerosols play a critically important role in the hydrological cycle affecting cloud formation and global and regional patterns of atmospheric circulation, such as the monsoon systems in tropical regions. They also have a direct effect on climate, by changing how much solar radiation is reflected or absorbed in the atmosphere. Humans change the aerosol loading by emitting atmospheric pollution (many pollutant gases condense into droplets and particles), and also through land-use change that increases the release of dust and smoke into the air. Shifts in climate regimes and monsoon systems have already been seen in highly polluted environments. A further reason for an aerosol boundary is that aerosols have adverse effects on many living organisms. Inhaling highly polluted air causes roughly 800,000 people to die prematurely each year. The toxicological and ecological effects of aerosols may thus relate to other Earth system thresholds. However, the behaviour of aerosols in the atmosphere is extremely complex, depending on their chemical composition, their geographical location and height in the atmosphere. While many relationships between aerosols, climate and ecosystems are well established, many causal links are yet to be determined. This means it has not been possible to set a specific threshold value at which global-scale effects will occur.

11.11   Chemical Pollution (e.g. Organic Pollutants, Radioactive Materials and Plastics)

Chemical pollutants undermine the integrity and health of animals, humans and ecosystems. Different chemicals harm different systems of the body in different ways. This can be compromising the immune system or causing an imbalance in sex hormones. As humans, animals and plants differ in their genetic makeup and also in their reaction to chemical dose, certain populations (plant animal and human) are more susceptible to toxic chemicals than others. Often exposure to a combination of chemicals can cause more problems than to a single toxic chemical.

Emissions of toxic compounds such as heavy metals, synthetic organic pollutants and radioactive materials, represent some of the key human-driven changes to the planetary environment. These compounds can persist in the environment for a very long time, and their effects are potentially irreversible. Even when the uptake and bioaccumulation of chemical pollution is at sublethal levels for organisms, the effects of reduced fertility and the potential of permanent genetic damage can have severe effects on ecosystems. For example, persistent organic compounds have caused dramatic reductions in bird populations and impaired reproduction and development in marine mammals. There are many examples of additive and synergic effects from these compounds, but these are still poorly understood scientifically. At present, we are unable to quantify the chemical pollution boundary, although the risk of crossing Earth system thresholds is considered sufficiently well-defined for it to be included in the list as a priority for further research.

Chapter 12
Climate Adaptation & RWH

12.1 Climate Adaption

Human society has been adapting to their environments throughout history by developing technology, cultures and livelihoods which are suited to local conditions. Climate change raises the possibility that existing societies will experience climatic shifts that previous experience has not prepared them. Climate adaption refers to the actions that society takes to combat the problems of climate change. These problems can include flooding, drought, storm events, sea level rise and land loss. Other problems are associated with fossil fuel use and include acidification of the oceans and loss of biodiversity. There are a wide variety of technological and social adaptation measures possible. Amongst these are green-grey infrastructure, sometimes referred to as hybrid infrastructure. Hybrid infrastructure is a combination of high tech human built engineering constructs (grey infrastructure) and nature-based solutions (green infrastructure). This hybrid approach can result in a more resilient infrastructure network which can adapt to the changing climate more efficiently than the current approach of grey engineering technologies. The concept of resilience being the ability of a system to resist shock, which is destructive, and through resilience move to a stress, which can be coped with, is fundamental to future infrastructure schemes. The emphasis in future schemes will be on locally available resources, decentralised systems and low embodied carbon and energy in construction, operation and maintenance. This chapter discusses how rwh can play an important role in these adaptions.

12.2 Design Water In

Over the past 50 years rainfall patterns have changed globally. The areas where rainfall levels and rainfall events have increased, in both volume and intensity respectively, have therefore problems with excess water. The concept of rwh, where rainfall is captured for storage and reuse, or diverted for use in irrigation or ground-water recharge, results in reduced rainfall entering watercourses and stormwater infrastructure. **The principle of designing water in rather than designing water out is relevant to adaptions to climate change.** Rainfall and local watercourses should be seen as assets, which can be designed to augment mains water supply, supply flood control and provide amenity and aesthetic value. The harvested rainfall is a resource and a potential source of potable water/irrigation water. But there is also the fact that rainfall captured, stored or diverted by this principle, is removed from the area threatened with flooding.

At the moment surface water infrastructure is designed to remove water from urban areas and expensive pipework is utilised to design water out of the area. This infrastructure has come under increasing strain in the last few years as it struggles to cope with water volumes way beyond the original water flows used as design calculations for the pipework. Therefore, rwh can serve to reduce the stress on surface water infrastructure during storm events. It does this both by storing or diverting water volumes. It can also, in extreme storm events, contribute to flood alleviation by slowing down the ingress of excess water into surface water infrastructure. The rain cities of South Korea discussed in an earlier chapter illustrate how this can be achieved in practice.

12.3   Centralised Versus Decentralised

The water infrastructure of both developed and developing countries have been designed and built as centralised structures where water is generally captured outside the point of use and delivered to urban areas where the major populations are. These centralised systems are based on previous rainfall statistics and have a lifespan of the order of 30 to 40 years. They have proved difficult to maintain, with leakages in and out of systems and have been seen as unsustainable in the concentration of embodied energy in the materials used in construction. As rainfall patterns continue to change, and as centralised systems become more difficult to maintain and less resilient to the effects of climate change, **there is a role for decentralised small-scale systems to be used in parallel with centralised systems**. This will allow for increased resilience as the smaller decentralised systems will have a smaller carbon footprint and will be capable of supplying ecosystem services. These are the benefits accruing to humankind from nature. These benefits include amenities for recreation through parks, waterways, and cycle paths but also areas to increase biodiversity and supply habitat rehabilitation.

12.4   Multiple Waters for Multiple Users for Multiple Uses

Rwh is integral to the new approach to water use and water treatment encapsulated in the phrase "**multiple waters for multiple users for multiple uses**". The concept of "multiple waters", where all waters, and their contents, are seen as a resource replaces the traditional concept of potable water, non-potable water and wastewater. Its basic tenet, that a water produced by one industrial process may be suitable, without treatment, for use in another industrial process changes the way we view water quality. If water produced by one factory is capable, without further treatment, of serving as a process water for a different factory, then the water is said to be "**fit for purpose**" and can be used without having to undergo expensive and unnecessary treatment. Other process waters from other industrial processes may contain minerals, nutrients heat or even water itself and these contents, can serve as source waters also. If society has multiple waters produced by industry then this water can be supplied to multiple industries i.e. multiple users, for multiple uses. In this scenario rwh is a viable and real alternative source of water.

12.5   RWH as a "Renewable" Water Supply

The viability of rwh as an alternative source of water supply can best be illustrated by considering the advances made in the fields of wind and solar energy in the last two decades. There is a globally recognised need to supply energy in a more sustainable way. Electricity production by burning fossil fuels is seen as contributing to many global problems and is due to be phased out in the coming decades. Investment in wind and solar technology is set to increase and targets have been set by the Intergovernmental Panel on Climate Change (IPCC) and other bodies that will see solar and wind technology increase the amount of electricity produced from these sources. However, it is not envisaged that wind and solar power sources will completely replace other sources of electricity. If the problem of storage of electricity from renewables is overcome perhaps this may occur in the future, but it is not anticipated at the present. In a similar way rwh can have an important role in augmenting existing water supplies, and investment should be encouraged to promote public and corporate acceptance and development of rwh as an important principle in water management. As a potential source of potable water, as a source of irrigation water or a source of recharge water for aquifers and groundwaters rwh is a technology that can serve society in becoming more sustainable and resilient.

12.6   Water and Agriculture

Rwh will continue to form an important contribution in agricultural water supply. For centuries rwh has served as a major source of water in most subsistence farming. Cultures around the world timed plant production to the seasons constructing labour intensive sustainable infrastructure to capture the rains and water crops. Social and religious rites were also involved in these community tasks. With the advent of the Anthropocene and industrial farming, agricultural water has become an issue in developed and developing countries. Since the widespread use of nutrients (phosphates and nitrates) in the farming process, the increase in world population and globalisation of food production corporate agricultural companies have expanded into developing countries using vast areas of land for food production. As a result, food production figures have increased and countries in Latin America and Africa produce food and other grown crops for export. **This has resulted in a huge increase in the amount of water used in agriculture.** The term agricultural water is something of a misnomer, for when we think of agriculture and water use we tend to imagine a subsistence farmer utilising rain water and surface water to irrigate his/her crops. This is far from the case. Agriculture is now a multinational industry with most crops grown for export. Water sources have been depleted, others contaminated, and the indigenous farmers displaced from their lands. Though rwh still plays an important part in subsistence farming, where it is often subjected to the vagaries of climate change, land rights issues and water rights issues have served to increase the marginalisation globally of subsistence farmers. This has added to the increased process of urbanisation in most countries as farmers find it increasingly more diffi-cult to eke out a basic living on the land.

12.7   RWH as a Leapfrog Technology

The technology underlying rwh which can be constructed using local resources, is ideally suited to decentralised applications in villages and small towns. It does not require the heavy infrastructure and treatment systems associated with large munic-ipal water projects, where dams or reservoirs are required as are pipe networks to deliver to households and industry. The water delivered in the pipework is treated water and this in turn requires infrastructure power and maintenance. Such water systems are capital intensive projects, requiring large investment and highly skilled engineering inputs. Such skills and finance are not readily available in developing countries nor are the local government structures required to maintain and adminis-ter the end product. Many African utilities find themselves locked in this downward spiral of deteriorating water assets and service with many families lacking basic water and sanitation infrastructure. The advantage of rwh technology using locally available resources is that developing countries could leapfrog the capital-intensive water infrastructure. The issue is similar to what occurred with mobile phones where the new mobile networks allowed developing countries to leapfrog over the expensive infrastructure required for telephone land lines. Thus, developing coun-tries went from no phone networks to extensive mobile networks while avoiding costly infrastructure. Rwh as a light, green creative solution could play a similar role in transforming water supply.

12.8   RWH as a Driver of the Bluewater Economy

In much the same way as mobile phone technology has had an impact on driving innovation in rural economies, rwh can also be used as a driver of the blue water economy. Rwh systems can be designed to be standardized, modular and plug and play. The philosophy is that all technological parts are standardized, i.e. of a speci-fied type and size and these standardized parts are based on local availability. Therefore, all parts of the systems will be available locally. The same parts can be used in all locations in a particular region. This will both create a market for these parts but will also ensure that these parts are available, and the technologies will continue to function. The technological systems themselves be they rainwater har-vesting systems or natural wastewater treatment systems are made up of different modules. This facilitates maintenance. If a module fails due to age or overuse, a replacement module can be installed without compromising the whole system. This modular system will also facilitate scaling individual household systems to village systems, if required. The concept of plug and play allows for the set-up of the technology and its use without user involvement in commissioning or designing the technological product. The individual components are simply put together, "plug" and are instantly fit for use "play".

12.9 Health Risks of RWH

The health risk associated with untreated water is frequently cited as a reason not to incorporate rwh as a water supply strategy. The concern with preserving the quality of treated water, potable in most cases, is a legitimate one. The main health risk associated with rwh technology is cross contamination of municipal and other treated potable water sources from rwh connections. Such concerns are groundless in situations where harvested rainwater serves as a source for irrigation water or groundwater recharge. In the case where harvested rainwater is to be used as a source for potable water, the technology exists, low tech and capable of being produced from locally available materials, to produce a high-quality water meeting potable and higher water standards. It is a design factor and if specified and properly installed and maintained filters and diversion systems are capable of meeting the highest water quality standards.

12.10 The Role of RWH in the SDGs and the Planetary Boundaries

In striving to combat the problems of the Anthropocene society has introduced the concepts of the Sustainable Development Goals (SDG's) and those of the Planetary Boundaries (PB's). Common to both systems is the aim to provide a guide to humans on how best to address the global problems we face. They promote a sustainable use of resources and point humanity toward a way of living that is aware of human behaviour and the effect this behaviour is having on all living and non-living things on the planet and its surroundings. Both emphasise the importance of the resources humanity require to exist, none more so that water. As W.H. Auden, the English poet said, "thousands have lived without love, none without water". In terms of utilising resources sustainably and living within the planetary boundaries rwh technology allows humanity to capture an alternative source of a resource and also to guard and channel this resource to the benefit of nature and the planet.

12.11   Peak Water

The term peak water is borrowed from the term peak oil, which sees the present extraction of oil as being greater than its replenishment, hence peak oil means that oil resources are now at their peak and in the future are set to decline. Peak water refers to the fact that the rate of water extraction globally is greater than ground and surface water replenishment. Further climate change has impacted on available freshwater supplies with receding glaciers, reduced river flow and shrinking lake volumes. Pollution due to contamination by human and industrial wastes has also impacted on water courses as has overuse of aquifers, particularly non-renewable groundwater aquifers. As a result in some regions the total freshwater supply has become polluted and unavailable for potable use, industrial use or even agricultural use. Presently, water demand exceeds supply in many parts of the world. These challenges, taken together with an increasing global population, has led some experts to talk about our present water resources as being at their peak and that the years ahead will lead to a decline in both water quality and water quantity. In this scenario the importance of rwh technology can be seen as it offers an alternative source of water that can serve as a source of potable water, irrigation water and also groundwater recharge.

12.12   The Circular Economy of Water

African society inherently understands the concept of circular economies at the community level. On a daily basis, communities extract the maximum value from each element of any supply chain, reusing and recycling resources to generate minimal waste. This is particularly true of water in rural sub-Saharan Africa for example, where water used for cleaning and bathing is not discarded but rather used for animals and irrigation. While the mindset for a circular water economy is present in Africa, other factors such as declining water supplies due to drought or poor-quality water due to pollution mean that water scarcity is still an issue. Rwh technologies can address such factors, increasing the efficiency of circular water economies, so that rural communities in developing countries can maintain a supply of suitable quality water that will meet local water demands. For example, rainwater harvesting systems can be used in conjunction with dedicated storage containers to collect rainwater during 'wet' seasons and store it for use in 'dry' seasons.

Finally, the human factor must also be included in any adaptions to climate change. The industrial revolution saw the start of a trend that has accelerated over the years where the focus of society has been on economic production and industrial growth. The role of the individual, be it in rural or urban areas was to facilitate this growth by supplying labour food or other goods. The type of society illustrated in the chapter on ancient water systems showed a balance between social interaction, the use of technology and the environment. This balance is not a feature of modern society where the environment is seen as something to be exploited for profit, technology runs everyday life and social interaction is everyday reduced. The potential of rwh technology, and the interaction of a sustainable population utilising a sustainable technology to generate a sustainably used resource should not be underestimated to regenerate the balance of nature, technology and humankind.

References

ABS (Australian Bureau of Statistics) (1994) Regional statistics, NSW. ABS, Canberra, ABS catalogue no 1304.1

Ahmed W, Gardner T, Toze S (2011) J Environ Qual 40(1):13–21

Borella P, Teresa Montagna M, Romano-Spica V et al (2004) Legionella infection risk from domestic hot water. Emerg Infect Dis 10(3):457–464

Brodribb R, Webster P, Farrell D (1995) Recurrent campylobacter fetus subspecies bacteraemia in a febrile neutropaenic patient linked to tankwater. Commun Dis Intell 19:312–313

Coombes PJ (2002) Rainwater tanks revisited: new opportunities for urban water cycle management. PhD thesis, University of Newcastle, Callaghan

Coombes PJ (2015) Discussion on the 'influence of roofing materials and lead flashing on rainwater tank contamination by metals' by M. I. Magyar, A. R. Ladson, C. Daiper and V. G Mitchell. Aust J Water Resour 19(1):86–90

Coombes PJ, Kuczera G, Kalma JD, Argue JR (2002) An evaluation of the benefits of source control measures at the regional scale. Urban Water 4(4):307–320

Coombes PJ, Dunstan RH, Spinks AT, Evans CA, Harrison TL (2006) Key messages from a decade of water quality research into roof collected rainwater supplies. Hydropolis conference, Burswood Entertainment Complex, Perth

Cunliffe DA (1998) Guidance of the use of rainwater tanks. National Environmental Health Forum Monographs. South Australian Health Commission, Adelaide, Water series 3

Dean J, Hunter PR (2012) Risk of gastrointestinal illness associated with the consumption of rainwater: a systematic review. Environ Sci Technol 46:2501–2507

Evans CA, Coombes PJ, Dunstan RH (2006) Wind, rain and bacteria: the effect of weather on the microbial composition of roof-harvested rainwater. Water Res 40(1):37–44

Evans CA, Coombes PJ, Dunstan RH, Harrison TL (2009) Extensive bacterial diversity indicates the potential operation of a dynamic micro-ecology within domestic rainwater storage systems. Sci Total Environ 407:5206–5215

Franklin LJ, Fielding JE, Gregory J et al (2009) An outbreak of Salmonella typhimurium 9 at a school camp linked to contamination of rainwater tanks. Epidemiol Infect 137:434–440

Fuller CA, Martin TJ, Walters RP (1981) Quality aspects of water stored in domestic rainwater tanks (a preliminary study). Domestic Rainwater Tanks Working Party, Australia

Heyworth JS, Glonek G, Maynard EJ, Baghurst PA, Finlay-Jones J (2006) Consumption of untreated tank rainwater and gastroenteritis among young children in South Australia. Int J Epidemiol 35(4):1051–1058

HSE (Health and Safety Executive) (2013) Code of practice: legionnaires' disease: the control of Legionella Bacteria in water systems, 4th edn. HSE Books, Sudbury

© Springer Nature Switzerland AG 2021
L. McCarton et al., *The Worth of Water*,
https://doi.org/10.1007/978-3-030-50605-6

Jin G, Jeng HW, Bradford H, Englande AJ (2004) Comparison of E. coli, enterococci, and fecal coliform as indicators for brackish water quality assessment. Water Environ Res 76(3):245–255

Kruse E, Wehner A, Wisplinghoff H (2015) Prevalence and distribution of Legionella spp in potable water systems in Germany, risk factors associated with contamination, and effectiveness of thermal disinfection. Am J Infect Control 44(4):470–474. https://doi.org/10.1016/j.ajic.2015.10.025

Kus B, Kandasamy J, Vigneswaran S, Shon HK (2010) Analysis of first flush to improve the water quality in rainwater tanks. Water Sci Technol 61(2):421–428

Luo C, Walk ST, Gordon DM et al (2011) Genome sequencing of environmental Escherichia coli expands understanding of the ecology and speciation of the model bacterial species. Proc Natl Acad Sci USA 108(17):7200–7205

Lye DJ (2009) Rooftop runoff as a source of contamination: a review. Sci Total Environ 407(21):5429–5434

McCarton L, O'Hogain S (2017) Thermal inactivation analysis of water related pathogens in domestic hot water systems. J Environ Eng Sci 12(2):34–41. Available at https://www.icevirtuallibrary.com/doi/10.1680/jenes.16.00028

Makin T (2014) Preheated water report: water regulations advisory scheme. See https://www.wras.co.uk/downloads/public_area/publications/general/preheated_water_Nov_2014.pdf. Accessed 20/09/2016

Martin A, Coombes PJ, Harrison TL, Dunstan RH (2010) Changes in abundance of heterotrophic and coliform bacteria resident in stored water bodies in relation to incoming bacterial loads following rain events. J Environ Monit 12(1):255–260

Mobbs M (1998) Sustainable house. Choice Books, Sydney

Morrow AC, Coombes PJ, Dunstan RH, Evans CA and Martin A (2007) Elements in tank water – comparisons with mains water & effects of locality & roofing materials. Rainwater and urban design conference 2007, Sydney, pp 830–837

Morrow A, Dunstan H, Coombes PJ (2010) Elemental composition at different points of the rainwater harvesting system. Sci Total Environ 408(20):4542–4548

Murrell WG, Stewart BJ (1983) Botulism in New South Wales, 1980–1981. Med J Aust 1(1):13–17

NHMRC (National Health and Medical Research Council) (1996) Australian drinking water guidelines. NHMRC, Commonwealth of Australia, Sydney

O'Hogain S, McCarton L, McIntyre N, Pender J, Reid A (2012) Physicochemical and microbiological quality of harvested rainwater from an agricultural installation in Ireland. Water Environ J Promot Sustain Solut 26(1):1–12

Prescott LM, Harley JP, Klein DA (1993) Microbiology. Wm. C. Brown, Dubuque

Rodrigo S, Sinclair M, Forbes A, Cunliffe D, Leder K (2011) Drinking rainwater: a double-blinded, randomized controlled study of water treatment filters and gastroenteritis incidence. Am J Public Health 101(5):842–847

Simmons G, Hope V, Lewis G, Whitmore J, Gao W (2001) Contamination of potable roof-collected rainwater in Auckland, New Zealand. Water Res 35(6):1518–1524

Spinks AT (2007) Water quality, incidental treatment train mechanisms and health risks associated with urban rainwater harvesting systems in Australia. PhD thesis, University of Newcastle, Callaghan. See http://urbanwatercyclesolutions.com. Accessed 09/05/2017

Spinks AT, Berghout B, Dunstan RH, Coombes PJ, Kuczera G (2005) Tank sludge as a sink for bacterial and heavy metals and its capacity for settlement, re-suspension and flocculation enhancement. 12th International rainwater catchment systems conference – mainstream rainwater harvesting, New Delhi

Spinks AT, Dunstan RH, Harrison T, Coombes P, Kuczera G (2006) Thermal inactivation of water-borne pathogenic and indicator bacteria at sub-boiling temperatures. Water Res 40(6):1326–1332

Weinstein P, Macaitis M, Walker C, Cameron S (1993) Cryptosporidial diarrhoea in South Australia. Med J Aust 158:117–119

WHO (World Health Organization) (2003) Heterotrophic plate counts and drinking water safety: the significance of HPCs for water quality and human health. IWA Publishing, London

Index